BREAKING AWAY from the MATH BOOK

Creative Projects for Grades K–6

BREAKING AWAY from the MATH BOOK

Creative Projects for Grades K–6

PATRICIA BAGGETT
ANDRZEJ EHRENFEUCHT

SCARECROWEDUCATION
LANHAM, MARYLAND • TORONTO • OXFORD

Published in the United States of America
by ScarecrowEducation
An imprint of The Rowman & Littlefield Publishing Group, Inc.
4501 Forbes Boulevard, Suite 200, Lanham, Maryland 20706
www.scarecroweducation.com

PO Box 317
Oxford
OX2 9RU, UK

The original edition of *Breaking Away from the Math Book* was published in 1995 by Technomic Publishing Company, Inc.

First ScarecrowEducation paperback edition 2004

Baggett, Patricia.
Breaking away from the math book : creative projects for grades k–6 / Patricia Baggett, Andrzej Ehrenfeucht.
p. cm.
Originally published: Lancaster, PA : Technomic Pub. Co., c1995.
ISBN 1-57886-159-4 (pbk. : alk. paper)
1. Mathematics—Study and teaching (Elementary) I. Ehrenfeucht, Andrzej. II. Title.
QA135.6.B34 2004
372.7—dc22

2004013015

The paper used in this publication meets the minimum requirements of American National Standard for Information Sciences—Permanence of Paper for Printed Library Materials, ANSI/NISO Z39.48-1992.
Manufactured in the United States of America.

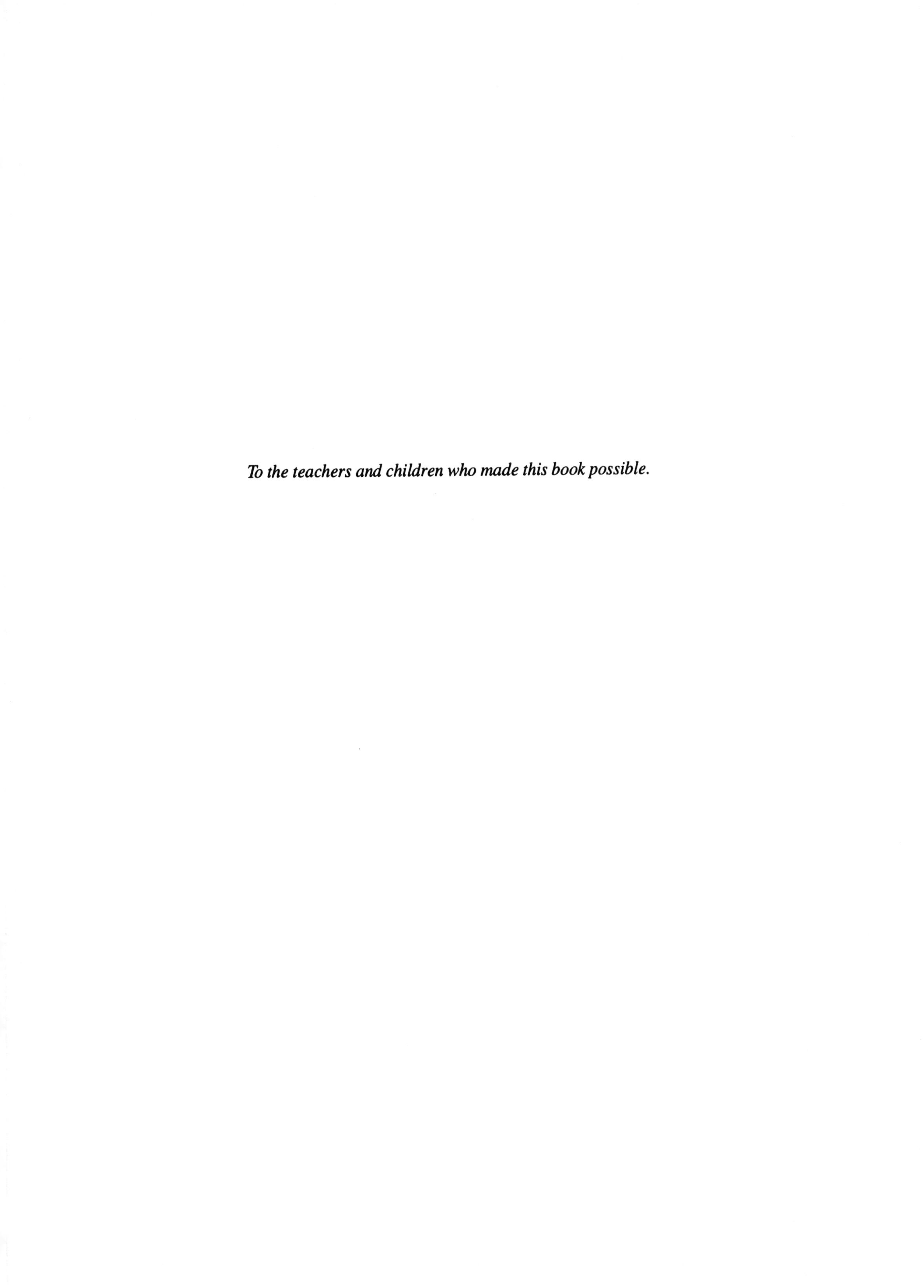

To the teachers and children who made this book possible.

TABLE OF CONTENTS

PREFACE

The National Council of Teachers of Mathematics recommended the use of calculators in classrooms in 1989. But in spite of this recommendation, children in early grades learn the same mathematics that their great-grandparents did more than 100 years ago. They practice laboriously the four basic operations on worksheets, learning how to add, subtract, multiply, and sometimes, how to divide, first small whole numbers, later common fractions, and finally decimals.

This way of teaching and learning has a long tradition and was well justified at the time when skill in paper and pencil computation was a necessary prerequisite to solving any problem that required mathematics. At present some skill in mental arithmetic, coupled with the right use of calculators, enables a person to solve problems that before required years of study.

It is not surprising that children find arithmetic boring. Their intellectual development is always ahead of their arithmetic skill. So in the second grade they work on problems that could be of interest to a preschooler. In the sixth grade they work on problems children in second grade might find interesting, and so on.

This book breaks from the tradition of paper and pencil computation. Children who learn some mental arithmetic and learn to use four-operation calculators are equipped to solve problems that interest them. They are limited not by their skill but only by their knowledge of the world and their understanding of mathematics. Children are curious, and when presented with problems that interest them, they show a willingness, persistence, and delight in learning.

Is breaking with the tradition of paper and pencil computation a radical change? No. The adult world has adapted to it. In the adult world, computations are performed mentally or with computing devices. We are simply bringing to children a change that has already occurred in the adults' world.

ACKNOWLEDGEMENTS

The lessons in this book have been taught by many teachers in the Jackson Intermediate School District and the Jackson Public Schools, Jackson County, Michigan. And many teachers have allowed the first author of this book to try out materials included here in their classrooms. Without the willingness of teachers, principals, administrators, parents, and children to break away from traditional elementary school mathematics, this book never could have come into existence. The administrators of the Jackson ISD (Dr. Mike Marlow, math/science coordinator, who encouraged us from the start; Dr. Jim Honchell, assistant superintendent; and Jane Weiser, kindergarten and gifted/talented coordinator) deserve a special thanks for allowing us to begin our math program in their district. Penny Vogel, a math specialist in the Jackson Public Schools, also aided us in beginning our program in the city of Jackson. The comments of the children who participated will never be forgotten. They've written letters, sent e-mail, lent artifacts that they have created during the lessons, and, in general, shown an enthusiasm for mathematics that is more than we thought was possible.

The schools, principals, and teachers involved have included:

- George Long Elementary School, Grass Lake, MI: Bob Bullock, principal; Marilyn Jabkiewicz, Kathy Holzer, Monica McGill, Cyndy Wallin, Suzie Macfarlane, Sharon Bechtel, Lana Freeman, Linda Bush, Sue Lawson, Darcel Hall
- Springport Elementary, Springport, MI: Sue Gallie, principal; Margie Mohr, Marilyn Sprinkle, and Karen Bradley (a special thanks to these three teachers, who have become "lead" teachers, helped with workshops, etc.); Sally Wurst, Jill Good, Heidi Johnson, Sally Wehrenberg, Nancy Sharrer, Mary Mahler, Sally Zimmerman, Barb Hoffman, Polly Pakonen, Todd Overweg, Janice Sanford, Denise Tomasello, Rex Carter, Pat Cleary, Shawn Williams, Sheryl Irvine, Tina Bienz, Dennis Bradley, Barb Hoag, Kathy Carpenter, Bob Millikin, Dave Lutzka, Gary Gardiner, Kathy Zinker, Pat Taber, Helen Schmidt, Bonnie Smock, Paula Ruhl
- Allen Elementary School, Jackson, MI: Linda Bryan, principal until 1994; Barbara Alexander, principal, 1994–1995; Mary O'Neal, Sharon Harper, Willie Pigott, Sarah Battles, Marilyn Scott, Mary Maddalena, Michelle Stocker, Linda McCully, Jane Schultz, Nancy Cudworth
- Parma Elementary, Parma, MI: Joyce Pickett, principal; Molly Frever, Joyce Wilkerson, Lyn Baker, Steve Bartels
- Mitchener Elementary, Adrian, MI: Kathy Blanchard, principal; Archie Handy

- Northwest Elementary, Jackson, MI: Dennis Demaris, principal; Cindie Sales, Jann Kilgore
- Western Middle School, Jackson, MI: Jim Warden, teacher consultant; Mike McKinney, Nancy Estes
- Townsend Elementary School, Jackson, MI: Marsha Bowers, principal; Colleen Stapley, Diane McQuillan, Sharon Elzinga, Nancy Frankila, Yvonne Rininger, Lisa Osypczuk, Charlene Roberts, Helen Kalinowski, Anita Johnson

We also thank Deborah M. Gibson, Mary Jane Northrup, and Roger Espinosa for creating the illustrations for the book.

INTRODUCTION

This book is intended for classroom teachers and parents who want to show their children that mathematics is both useful and interesting. It has been designed so that the user has ample space for writing notes in the margins.

In this book you will find material for forty-five lessons in mathematics, and a tutorial on how to use a calculator (Chapter 46). The lessons are written in two very different styles. One style, shown for example in ''Traveling Bugs'' (page 69), is very brief, almost telegraphic. The other, shown for example in ''Exploration of Stars'' (page 147), is a narrative. Each narrative is a description of a lesson taught by the first author as a guest instructor in a classroom. Such narratives describe not only the content of the lesson, but also the teaching style of the instructor and the comments and actions of the children. The short descriptions give only a plan for a lesson, and they usually require a considerable amount of additional preparation. The teachers who have used these materials have preferred the narratives at first; only after gaining considerable experience with the material have they found the short versions adequate.

All of the lessons in this book have been tested in classrooms by teachers and by the first author, and in workshops for teachers and parents. Most have been tested a number of times, and some are regularly used by teachers who include them in their standard curriculum. A number of parents have also used them on a one-to-one basis with their children, as fun and challenging activities at home. To do the projects at home, each child should have a four-operation calculator; in school the teacher should have an overhead projector, a transparent overhead calculator, and a calculator for each child.

All sequences of keystrokes were tested on a TI-108 calculator, produced by Texas Instruments, but they also work, sometimes with minor adjustments, on most generally available brands. In a classroom it is important that all children use the same brand and model of calculator. Special calculators, such as the TI Explorer (which uses common fractions), and scientific or graphing calculators, are not only unnecessary, but are inappropriate for projects in this book.

Two questions are often asked. Which lessons are for which grade? And, how should you combine them with the existing curriculum? For many lessons we have indicated in which grades they have been taught. Most have been taught in grades two to six. Some, such as ''Store for Squirrels,'' and ''When I went to St. Ives,'' were especially prepared for kindergarten and first grade and would be boring for older children. ''Area of a Triangle (Heron's Formula)'' and ''Leaves'' are also appropriate for middle school and even high school. ''What is Fair?'' created a very interesting discussion about right and wrong in one seventh-grade class. Generally, the lessons progress from easy at the beginning of the book to difficult at the end, but this is not an absolute rule. Many lessons can be taught at very different levels. For example, "Computing Lengths of Diagonals with the Pythagorean Theorem" has been taught in grades two through eight.

As long as children are given standardized tests, they must be prepared for them. So, as long as paper and pencil subtraction is one of the items on the tests, children have to learn it. This means that material from this book cannot be the only mathematics that children study, and it sometimes leads to some paradoxical situations. One day children in a second-grade class had the "Shop 'n' Save" lesson, which requires multiplication of decimals, and the next day they did worksheets subtracting two-digit whole numbers without regrouping. Teachers who have spent one or two periods per week on material from this book and the rest of their available math time on their standard curriculum report to us that the smaller amount of time on "regular math" was compensated for by the children's increased interest in math and their better performance.

Below we give some general suggestions about using the material from this book. The suggestions are based on teachers' observations and the results of tests given to the children.

(1) Children use tools such as rulers, protractors, and compasses extensively in the lessons. They should learn to use such tools properly and skillfully, and this takes time. Work with tools should be carried out until children achieve skill in their use.

(2) In early mathematics (when children learn counting), manipulatives of various sorts are absolutely essential, and they cannot be replaced by counting pictured objects.

(3) Children should make measurements themselves, so they know how the numbers in the problems were created. Young children rarely assign any real meaning to numbers that are provided by the teacher or read from the book.

(4) Most lessons include several different activities, such as talking, reading, measuring, and interpreting numbers. Often an essential part of a lesson is recording the numbers in a useful form. This typically involves both preparing a data sheet and filling it in. Learning to prepare a data sheet (and not using a premade worksheet) is an important part of many lessons.

(5) Performing calculations is often only a small part of a problem—and not the most important.

(6) A lesson may last thirty minutes or may extend over several days and several sessions. The same topic (e.g., "Menu") can be used in several consecutive lessons, with activities involving different organizations of calculations. Examples are: using the calculator's memory or not using it, learning to make change, and finding the cost of meals in the school cafeteria. But after a topic is closed, we do not encourage going back to it even if some children "didn't get it." Very definitely we discourage going back to the topic with the same materials or the same manipulatives. This also includes remedial teaching. We suggest that remedial teaching be done with new material and not with material that has been used before.

(7) In learning to use a calculator, children should be encouraged to experiment with the keys and shown how to use all features of the calculator. To do this, a teacher can use specific activities such as games, unusual calculations, and interesting key combinations (such as [2][*][=][√]).[1] Such activities should be included regularly and emphasized until the children are really familiar with the calculator's features.

(8) Drill with calculators, such as adding groups of numbers or multiplying two integers, is not needed or encouraged. We know from experience that children acquire adequate calculator skills in the course of solving the problems suggested in the lessons.

[1]In this book, multiplication is shown by [*] and sometimes by [×].

CHAPTER 1

Learning to Use the Calculator

This introduction to the calculator has been taught in several third-grade classrooms. It takes about forty-five minutes. Variations of this lesson have been taught in second and fourth grades as well.

PROPS

- a calculator projected via an overhead projector, or a large calculator poster visible to all children (The best situation is to have both.)
- a calculator for each child and the teacher

THE LESSON

"Can you open your calculator? Do you see how to slide it out of its case? Then you can turn the case over and slide the calculator back into it. The round corners of the case match the round corners of the calculator—there is a right-side up and an upside-down!

"How do you turn the calculator on? Punch the on/clear button." (The teacher pointed to the button on the poster and pressed the [ON/C] button on the overhead calculator.)

"What do you see on the display when you turn the calculator on? A zero. Anything else? Yes, a decimal point to the right of the zero. Do you know what a decimal point is? We use it, for example, when we talk about money, to separate the number of dollars from the number of cents.

"What powers the calculator? The calculator has a solar cell. It needs light in order to work, so it won't work in a dark restaurant. If you put your finger over the solar cell" (she pointed to the solar cell on the poster) "and just wait a few seconds, what happens? The 0. begins to fade! Then just remove your finger, and the 0. will come back.

"How do you turn off the calculator? You don't! It will turn itself off if you don't press any keys for a while, or if you put it back into its case.

"What is the biggest number you can put into your calculator? Just start punching nines and see. You can keep punching; too many punches don't hurt. How many nines? Eight! Can anybody read that number? 99,999,999. That is ninety-nine million nine hundred ninety-nine thousand nine hundred ninety-nine! That is a lot!

"How do we know it's the biggest number the calculator will hold? Let's add one to it and see what happens. Does everyone have 99999999. on the display? Now press

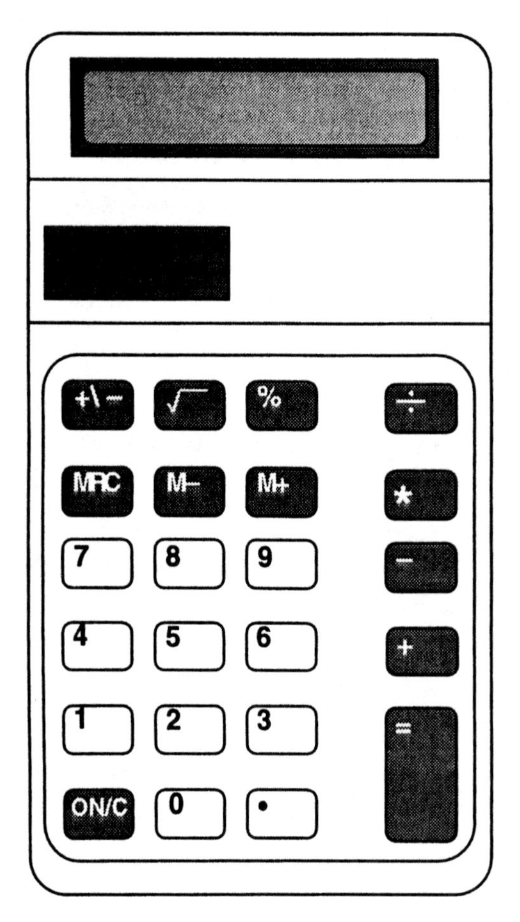

ILLUSTRATION 1. A four-operation calculator.

[+][1][=].'' She demonstrated on the poster and pressed the keys on the calculator. ''What do you get on your display? *E* 1. Uh oh! What do you think *E* means? Error! And now, look: your calculator is frozen. Just try to punch in another number. Nothing goes in. Have we broken our calculators? No! Just punch on/clear ([ON/C]), and your calculator becomes unfrozen. It is not broken. There are some other ways to get errors too, but we will not discuss them now. Just remember when you see *E* all you need to do is press [ON/C] to get your calculator to work again.

''How to add two numbers? Let's add 3 plus 5. Here are the keystrokes: [3][+][5][=]. What do you see? 8. What is 3 plus 5? 8.

''Now let's play a little game. I will call it the first-second game. I will put a number on the board and call it the first number. My first number is 5. You need to get 5 on your display. Now I give you my second number: 10. The game is to change the 5 into 10. The rule is, you can't just hit on/clear ([ON/C]) and enter 10. You have to do something to 5 to change it into 10. What can you do?'' Here suggestions from children can be tried. The most typical one is [+][5][=]. Each time a solution is suggested, the teacher should demonstrate it.

''OK, now we have 10. Let's undo it. Who can get us back to 5? We have to do something to 10 to get it back to 5.'' A common solution is [−][5][=]. ''OK, let's try it again. My first number is still 5, and my second one 10. But now let's pretend the plus key on the calculator is broken. How can I change 5 into 10 without using plus?'' A common solution is [*][2][=]. ''OK, now let's change 10 back into 5, but let's pretend the minus key is broken. How can

I change 10 into 5 without using minus? Do you know about division? [10][÷][2][=]. What do you get? 5!

"Now you can suggest some first/second numbers. They can be big numbers if you want." Here she took some number pairs from the children. Sometimes, when they were not difficult enough, the teacher suggested slightly different ones.

"How about 123 as a first number and 163 as a second number? What do we do? We need to change the 2 into a 6. How to do it? Add 40! Keystrokes: [+][40][=].

"Let's pretend we are going out for lunch, and I am going to buy eight of you a slice of pizza for $1.19 and twelve of you a hamburger for $1.35 and two of you a cheeseburger for $1.49. How much will I pay? (She wrote these prices and the number of children on the board.)

"For the pizza, I will pay 8 times $1.19. How do I get that on my calculator? Here are the keystrokes: [8][*][1.19][=].

"Did you get the decimal point between the dollar and the nineteen cents? What is the total? 9.52. What does it mean? I will pay $9.52 for the pizzas. OK, does everyone have 9.52 on the display? Now let's put it into the memory basket: Press the M-plus key ([M+]). What do you see on your display now? An *M*. What does that mean? The calculator has a place which we call memory, or the memory basket, where it can put a number. When the memory basket has a number in it, you see an *M* on your display.

"Now let's figure out how much I pay for the hamburgers. Here are the keystrokes: [12][*][1.35][=]. What do you get? 16.2. What does that mean? Sixteen dollars and two dimes, or sixteen dollars and twenty cents. The calculator says, 'I do not need to show the number of pennies if the number of pennies is zero.' Now we want to put this into the memory basket also. So press [M+]. How much should we have in memory now? $9.52 plus $16.20. How much is that? If you want to know, just press [MRC]. It is read Memory Recall. It will display for you the number which is in memory. What is it? 25.72. So far my bill, for eight slices of pizza and twelve hamburgers, is $25.72. Now let's figure how much I pay for two cheeseburgers at $1.49 each. [2][*][1.49][=]. What do you get? 2.98. Two dollars and ninety-eight cents. Let's put it in memory: [M+]. Now, how much is my whole bill? Press [MRC] to see. 28.7. That means twenty-eight dollars and seven dimes, or $28.70. Do you see that you still have an *M* on your display? That means that 28.7 is still in the memory basket. If you want to empty the memory basket, you need to press [MRC] twice in a row. You already pressed it once, so do it again now. The *M* disappears. So even though you see 28.7 on the display, it is no longer in the memory basket.

"Have you ever gone out to dinner, and when your parents paid the check, they left a tip? Let's say I want to leave a tip with this bill for $28.70. How much should I leave? Sometimes people leave a 15-percent tip. We have a percent key on our calculator: [%]. Let me show you how to calculate how much money I pay if I want to leave a 15-percent tip along with my $28.70. Do you have 28.7 on your display? Now press [+][15][%]. What do you see? 33.005. What does it mean? Thirty-three dollars and no dimes and no whole pennies and one-half of a penny. So how much money should I leave? I can't leave a half a penny, so I would probably leave thirty-three dollars.

"Here is a key: [√] Let's see if we can figure out what it does. Try these keystrokes." (She wrote on the board: [25] [√].) "What do you see? 5."

Some more:

[16][√]	Display: 4
[64][√]	Display: 8

[81][√] Display: 9
[4][√] Display: 2

"What is the key doing? Can anyone tell me? Let me show you something. [25][√] Display: 5. How do I change 5 back into 25? [*][5][=] Display: 25. You can use a shortcut here, and just press [*][=] to get back to 25. Try it! [16][√] Display: 4. How to change 4 back into 16? [*][4][=] Again you can use a shortcut: just [*][=]."

[64][√] Display: 8. How to change 8 back into 64? [*][=]
[81][√] Display: 9. How to change 9 back into 81? [*][=]
[4][√] Display: 2. How to change 2 back into 4? [*][=]

"What does the key do? It gives a number, and when you multiply that number by itself, you get the original number back! The key is called the square root key. So 5 is the square root of 25.

"Now try [2][√]. Display: 1.4142135. Will [*][=] get you back to 2? Try it. Display: 1.9999998. That is very close to 2, but it is not 2. Some numbers have square roots with lots of numbers after the decimal point.

"Would you like for your calculator to count by twos?

Try [+][2][=][=][=][=]. What do you see? 2, 4, 6, 8, . . .

Now make your calculator count by fives: [+][5][=][=][=][=] . . .

Suppose you want it to count by fives starting at 1.

So you want it to show 1, 6, 11, 16, 21, . . .

How to do it? [1][+][5][=][=][=][=]. Try it!

How can you get your calculator to count by halves?

It would show 1/2, 1, 1-1/2, 2, 2-1/2, 3, . . .

How does the calculator show 1/2? 0.5. Press [.5][+][=][=][=][=] . . .

What do you see? .5, 1., 1.5, 2., 2.5, 3., 3.5, . . .

"Can you get your calculator to count backward by ones from 10? [10][−][1][=][=][=][=]. . . . Try it and count with me: 10, 9, 8, 7, 6, 5, 4, 3, 2, 1, 0, −1, −2." (Children had difficulty here.)

"Do you know about numbers below zero? They are called negative numbers, and we see them sometimes on a thermometer when the temperature goes below zero. We need them sometimes when we spend more money than we have! There is a key on our calculator, [+/−]. It changes a number into a negative number, and if you press it again, it changes it back again. Try [5][+/−][+/−][+/−][+/−]. Do you get 5, −5, 5, −5? See how the calculator puts the minus sign away from the number, where the *E* and *M* have gone before?

"It is time to quit. But before I go, can you tell me again the name of this key: [√] ? It is square root, of course!"

CHAPTER 2

Some Beginning Activities

STAGE 1. PLAYING

Show children about the calculator: how to turn it on, that it has a solar cell, what to do when it freezes (press [ON/C]). (See Chapter 46, "Getting Familiar with the Calculator Keys.") Let them just play with the calculators. They will have lots of questions!

STAGE 2. STRUCTURED PLAY

Suggest some patterns for the children to enter, for example, all ones, or press the two key once and the zero key twice. Let them get familiar with the names of keys and how to read single-digit numbers. Working in groups, one child can name a number, and the other(s) can enter it on their calculators.

STAGE 3. RELATING CALCULATOR ACTIVITIES WITH ARITHMETIC AND THE WORLD

Ask children to enter numbers on the calculator that have meaning to them, such as how old they are, how many brothers and sisters they have, the day of the month.

Give children some objects to count. Show them how to count with the calculator using the keystrokes: [1][+][=][=][=][=]. Have them press the equals key as many times as they have objects. Let them read the numbers from the calculator. Does the number of objects they have match the number on the calculator?

Do some simple first-number, second-number (or given/target) games.

Examples

Put 1 on your calculator display.

Now change it to 2. How can you do it? You are not allowed just to press the on-clear key and then press the two key.

Change 2 to 4. How can you do it?

Change 4 to 5.

Keep the range of numbers within the children's knowledge.

Play do and undo:

Change 1 to 2. Now change 2 back to 1. Undo it!

Let the children choose the number pairs.

CHAPTER 3

Even and Odd Integers

INTRODUCTION

Checking if an integer n is even or odd is very simple with a calculator. Compute [n][÷][2][=]. The display shows $n/2$ in decimal notation. So if it ends with .5 the number n is odd; if there is nothing after the decimal point, the number n is even. (If you see anything else after the decimal point, either you did not start with an integer or you did not divide by 2.)

Examples

[17][÷][2][=]	Display: 8.5	17 is odd
[24][÷][2][=]	Display: 12	24 is even
[7.2][÷][2][=]	Display: 3.6	7.2 is not an integer
[17][÷][3][=]	Display: 5.6666666	The number was not divided by 2

THE LESSON

Sharing Cookies

Children work in groups. Each group gets a lump of clay, two large paper plates, and plastic knives. When three children are in a group, one is the baker, and two are customers who will share their cookies. When four are in a group, one is the sales clerk, one is the baker, and two are customers who share the cookies.

Making Cookies

Each group makes clay cookies (at least twenty flat disks, approximately one inch in diameter) and puts them on the baker's plate.

How Do Two People Share Some Number of Cookies?

(This part should be repeated with bigger and bigger numbers.)

Take six cookies from the baker's plate and put them on the customers' plate. Each customer takes three.

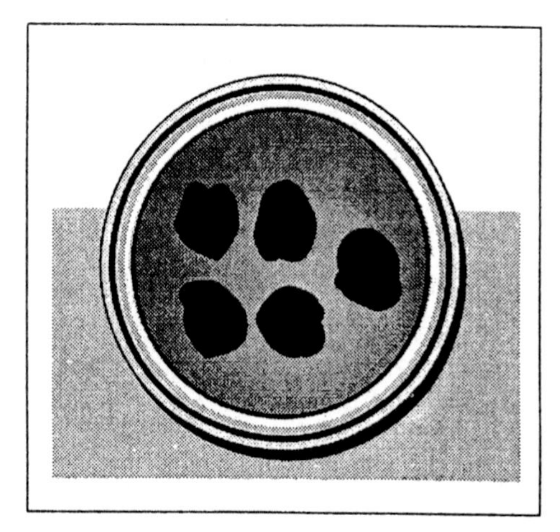

ILLUSTRATION 2. [5][÷][2][=] 2.5, so 5 is odd.

Take seven cookies from the plate. Each customer gets three whole cookies. We cut the remaining one in half, so each customer will get three and a half cookies.

Terminology: If we do not have to cut a cookie, the number of cookies is even. If we have to cut a cookie, the number of cookies is odd.

Guessing If a Number is Odd or Even

The activity remains the same, but before taking cookies from the plate, children guess if the number is odd or even, and how many cookies each customer gets. Take fifteen cookies from the plate. How many cookies for each person? Is fifteen even or odd? (Seven and a half, so fifteen is odd.) Now check.

Predicting with a Calculator How Many Cookies Each Person Will Get

When you press the divide key, the two key, and the equals key, the calculator divides the number by 2. If the number is odd, the calculator shows a half as .5.

EXAMPLE

[15][÷][2][=] Display: 7.5

Each person gets seven and a half cookies. The number fifteen is odd. Children still carry out the same activity, but instead of guessing, they predict with the calculator. Checking the predictions by dividing cookies into two groups is still important.

CHAPTER 4

Store for Squirrels

This lesson has been taught in several first-grade classes. The children had used calculators before but were not very experienced with them. Children worked in pairs; in each pair, one child was a porcupine and the other a squirrel.

PROPS

Each porcupine gets

- a zip-lock plastic bag containing four pecans, six walnuts, six buckeyes (horse chestnuts), twelve almonds, and fifteen hazelnuts (All are in their shells. Do not substitute peanuts or pistachios, as they will be eaten!)
- a paper plate to hold the nuts
- a bowl to use as a cash register
- a list of prices on a sheet of paper (see the end of the chapter)
- a calculator

Each squirrel gets

- a cup with twenty pennies for buying groceries
- a bowl to hold the groceries bought (We used red plastic bowls for squirrels and white styrofoam ones for porcupines. It is a good idea to have the bowls be of two different colors.)
- a calculator

It is good to have on hand a few extra nuts of each type and a few extra pennies, in case of loss. Also some squirrels requested a price list of their own.

THE LESSON

"We are going to work in groups of two, so let's push desks together in pairs. Now either you or your partner needs to be a porcupine, and the other person should be a squirrel. Will the two of you please decide who is who? Now, I have lots of things to pass out. Some go to squirrels and some to porcupines. As I come around, you need to tell me which you are."

I handed out the bag of nuts, plate, bowl, list, and calculator to the porcupines, and the cup with pennies, bowl, and calculator to the squirrels. The children also had pencils, which some used to draw pictures of nuts.

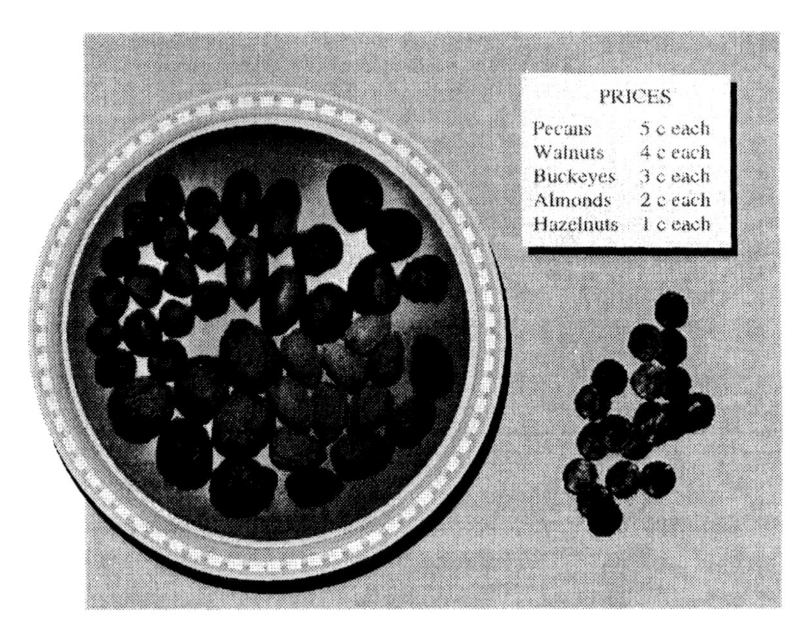

ILLUSTRATION 3. Some props for Store for Squirrels.

"Do you know what squirrels eat? One thing they eat is nuts. So the porcupines in the forest have opened some stores for squirrels, to sell them some groceries, which are nuts. Squirrels, you are going to get to buy some nuts. But first let's see what the porcupines have in their stores. Porcupines, will you dump out your nuts from your bag onto your plate? You want the goods in your store to look neat, so now sort your nuts so that all of one kind are together."

This activity created some excitement. There were a lot of nuts, and they almost filled the plate. The children sorted. Some could do it quickly and others took a long time.

"When you get the nuts sorted, put your price list up in front. I will write the prices on the board."

Price List	
Pecans	5¢ each
Walnuts	4¢ each
Buckeyes	3¢ each
Almonds	2¢ each
Hazelnuts	1¢ each

"Let's see if we can identify the nuts. Can anyone name any of the nuts in the porcupines' stores? Do you know what they are called? If you do, hold up the nut and tell me its name."

To my surprise, the children could name only one nut: walnut. They could not really read the word walnut, although they knew which sound to associate with the letter *w*. So the teacher drew a picture of a walnut next to its name on the board. Some children did the same on their price lists.

Eventually one child was able to name the (horse) chestnut. But since on the price list its name was buckeye, I explained: "Will each of you please get a horse chestnut? It has another name too, the buckeye. Where am I holding the chestnut?" I held it in front of my eye. "If you look at the horse chestnut, you will see that part of it is a soft brown color, while the rest is dark brown. The soft brown part is the shape of an eye, and it looks like the eye of a male deer. Do you know what a male deer is called? A buck. So it has the name buckeye!" The teacher drew a picture of a buckeye next to its name.

"What about the nut that has little dots on it? Do you know its name?" Silence. "It's an almond. Do you know any food that has almonds in it?" One child said, "Almond Joy!" The teacher drew a picture of an almond next to the word.

"How about this nut?" I held up a pecan. "Maybe you have had a pie made out of this nut. It is a pecan. Have you ever tasted pecan pie?" Some children had. "Let's draw a picture of a pecan next to its name.

"There is only one more kind of nut. Can you hold it up? What is it called?" Some children guessed acorn. "It does look a lot like an acorn, but it is not. It is called a hazelnut. Let's draw a picture of it by its name.

"Now we have the drawings on our price list. Who can read the whole price list to me now? Let's read it a line at a time. Who will read the top line?" Children struggled here, but eventually we got the whole thing read. They had to learn to read "¢" as cents. All children could read the numbers 5, 4, 3, 2, and 1.

"OK, we are finally ready to play store. Porcupines, are you ready? Do you have your price list out so that you and the squirrel can both see it?" ("Yes!") "Squirrels, do you have your money and your bowl, which will be your your grocery bag?" ("Yes!") "Squirrels, how much money do you have? Let's dump the money into our grocery bags and count it." They did, and we discovered that each squirrel had twenty cents.

"I will be the first squirrel. I would like to buy one walnut and two almonds. Squirrels, will you take one walnut and two almonds from the porcupine's plate and put them in your bowl?" This caused some commotion, as children were still not sure which nuts went with which names. "OK, here is the hard part. Now we squirrels need to pay. How much should we pay?"

The first guess I heard was "three cents, since there are three nuts." But most children thought it would be more than that. I asked if we could use our calculators to find out. We got them out and turned them on by pressing [ON/C].

"What keys do I need to press to figure out how much one walnut and two almonds cost?" Children said one walnut was four cents, so I should put in four. I did so (on an overhead calculator) and asked the children to do it too. "What next?" I asked. This was not so easy. One child said, "An almond costs 2¢." But it was not so obvious that we needed to add the 2¢. Eventually a child said we could "plus 2 cents for an almond." On my calculator I pressed [+][2]. "What else do I need?" I asked. Several children said, "Equals." I showed the class where the equals sign is on the calculator, and I pressed it: [=]. The display then showed 6. "Am I done?" I asked. "Is that all of my bill? Do I owe the porcupine six? Six what?"

"Six pennies," they replied.

"Is that all?"

A child finally piped up, "No. You need to pay for one more almond."

"How much is an almond?" I asked.

"Two cents."

"OK, what do I do on my calculator?"

"Plus two more cents," said several. I pressed [+][2][=]. The display showed 8.

"What do I have now?"

"Eight cents! That is the whole bill!"

"Does everyone agree?" I asked. "I will write on the board: [4][+][2][+][2][=]. Can anyone tell me why I did that?"

Several knew and could explain, "You pay 4¢ for a walnut and 2¢ for one almond and 2¢ for another almond."

"OK squirrels, will you please pay the porcupines? Put the right amount of money in the porcupine's cash register." The squirrels were excited to do this. There was lots of pouring the coins out on desks and counting. They knew they needed eight pennies. Eventually each squirrel in the class had put eight pennies in the porcupine's cash register. Finally the purchase of a walnut and two almonds was complete.

"Now, squirrels, how much money do we have left?" They counted and told me there were 12¢ left. "So what else should we buy? Let's spend all our money. Who would like to suggest what we can buy?"

Danielle said, "I want to buy a pecan."

"Squirrels, will you ask the porcupines to give you a pecan? Do you know which nut the pecan is?" Danielle got one from her porcupine and held it up. "Porcupines, please put a pecan in the squirrel's grocery bag. And squirrels, please pay. How much do you pay?" ("Five cents.") "All right, give the porcupine five cents." The children counted out their money. "How much money do you have left?" They counted. ("Seven cents.") "OK, what shall we buy now? Who can place an order that will take all the money?"

"I can," said Steven. "I will buy seven of the nuts that cost one cent each. But I don't remember what they are called!" One child said they were hazelnuts.

"OK, porcupines, give the squirrels seven hazelnuts. And squirrels, give the porcupines seven pennies. How much money do we have left?"

"Nothing!"

"Now let's switch who is who. Porcupines, you will now be squirrels, and squirrels, you will be porcupines. Dump all the nuts back in the plate and all the money back in the cup. Porcupines, make your groceries neat by sorting them."

"Shall we play one more time together, before you just play in pairs?" ("Yes!") "OK, this time let's plan our purchases before we make them. We have to be sure not to spend more money than we have. I want to buy just pecans. How many can I buy?" This question created some guessing. One child said three. "How much do I pay for three?" I asked. Silence. "Let's use our calculators. How much is one pecan?" ("Five cents.") "So what keys do I press on my calculator to see how much I owe?"

Suellen said, "[5][+][5][+][5][=]."

"Let's try it and see what we get." I showed the keystrokes on the overhead. 15! "How much do I have left?" This was tricky for the children, so I asked, "How much did I have to start with?" ("Twenty cents.") "And how much do I spend for three pecans?" ("fifteen cents.") "So how much should I have left?" Silence. "I will show you how to figure it out on your calculator. Do this: [20][−][15][=]. [−] means 'take away.' What do you see?" ("Five.") "What does it mean?" I could tell that not many children were following, so I asked the porcupines to give three nuts to each squirrel, and the squirrels to pay five cents for each of them, and to tell me how much money they had left. This was not so difficult, and pretty soon every child told me there were five pennies left. I wrote the program [20][−][5][−][5][−][5][=] on the board and explained, as I pressed the keys, "I started with twenty, and I paid five cents for the first pecan (so I take away five) and five cents for the second pecan (so I take away five cents more) and five cents for the third pecan (so I take away five cents more). Now I press 'equals,' and what does it tell me?"

Finally several children said, "How much you have left!"

"And how much is that?" ("Five cents.") "So can I buy another pecan?"

"Yes! And then you won't have any money left."

"So how many pecans can I buy for twenty cents?" ("Four.") "I want to show you something neat. Try this on your calculator: I bought four pecans, and each one costs five cents. You can compute the bill in another way, by using a new key, times: [4][*][5][=] (four times five equals). What do you get?" ("Twenty.") "OK porcupines, give your squirrels one more pecan, and squirrels, pay for the pecan."

Now it was time for the children to play in pairs. The children changed roles, and they went at it. I told them that each time after the squirrels bought the groceries they wanted, they should switch again, so within the class period each child got several opportunities to be a squirrel and a porcupine. They were still having difficulty naming the nuts, and they would frequently bring a nut to me and ask me what it was called and how much it cost. We heard snippets of conversations from the children: ''I won't give you the hazelnut until you give me the penny!'' ''Oh, I am spending too much. Can I trade in a pecan and get five cents back?'' When they had difficulty determining how much the squirrels needed to pay, I encouraged them to use their calculators. But after a while many didn't even need to use them. The teacher and I circulated around to help out when children asked.

The lesson was rather lengthy, and even when the bell rang, the children asked to play ''one more time.'' At one point I showed a pair of children how to use the [M+] key. For example, if the squirrel buys two walnuts and three buckeyes, these keystrokes will compute the total price: [2][*][4][M+] [3][*][3][M+][MRC].

We finally had to stop. The porcupines put their nuts in the zip-lock bags, and all supplies were collected.

SOME POSSIBLE CHANGES IN THE LESSON

(1) Since the children had difficulty the second time through with planning their grocery buying beforehand, I would first play two or three times without planning.

(2) I had prepared a ''New Prices'' list:

Pecans	5¢ each
Walnuts	three for 10¢
Buckeyes	two for 5¢
Almonds	three for 5¢
Hazelnuts	1¢ each

It seemed rather advanced for the first lesson, but I would have liked to try it in a later lesson, saying that the porcupine decided to have a sale.

(3) For older classes, I would give more money (e.g., one dollar in change) and make the nuts more expensive. For example:

Pecans	15¢ each or two for 25¢
Walnuts	10¢ each
Buckeyes	8¢ each or two for 15¢
Almonds	5¢ each
Hazelnuts	4¢ each or three for 10¢

(4) Making change is another topic that can enter in more advanced lessons.

(5) Squirrels could put the amount of money they have brought to the store in memory, then use [M−] and [MRC] as they make purchases, to keep track of how much they have left. For example, in buying one walnut and two almonds:

[20][M+]	
[4][M−][MRC]	Display: 16. I have 16¢ left after buying one walnut
[2][M−][MRC]	Display: 14. I have 14¢ left after buying one almond
[2][M−][MRC]	Display: 12. I have 12¢ left after buying another almond

(6) Repeated addition should also be included. For example, a hazelnut, two pecans, and four almonds cost [1][+][5][=][=][+][2][=][=][=][=]. (Display: 19.) Reentering numbers can cause keystroke errors, and with repeated addition you do not have to reenter a number.

PRICE LIST

Pecans	5 ¢ each
Walnuts	4 ¢ each
Buckeyes	3 ¢ each
Almonds	2 ¢ each
Hazelnuts	1 ¢ each

CHAPTER 5

When I Went to St. Ives

Once upon a time, in the city of Trento, in Italy, lived three sisters.
Each sister had four cats.
Each cat caught five mice.
How many mice were caught by the sisters' cats in the city of Trento in Italy?

This lesson has been taught in several kindergarten and first grade classes.

PROPS

- a sheet containing the story about three sisters' cats, a pencil, and a calculator for each child
- a transparency containing the story (see end of chapter) and an overhead calculator for the teacher

THE LESSON

"Can you help me solve a problem? You have a sheet that has a little story on it, and there is a problem in the story. I want to know the answer to the problem.

"Let's start by reading the story. Who can read the title? Margie?" Margie read, "When I went to St. Ives." She did not know how to read "St." or "Ives."

"Jerome, will you read the next line?" He read, "Once upon a time, in the city of Trento, in Italy, lived three sisters."

"Lacey, can you read the next line?" Lacey read, "Each sister had four cats."

"Amy, will you read the next one?" She read, "Each cat caught five mice." (Amy could not read the word "caught.")

"George, will you read the last line?" George read, "How many mice were caught by the sisters' cats in the city of Trento in Italy?"

"OK, we have a problem. Let's see if we can say again what we want to know. Who can tell me? Raise your hand if you know." Lots of hands went up, and children were already beginning to draw on their sheets.

"We want to know how many mice. And we can draw it and count," said Leslie.

"How do we draw it? I don't get it," said Julian.

"Who can suggest what we should draw?" I asked.

“I think we can draw the sisters,” said Andrew.

“How many sisters?” I asked.

“Three!” said Andrew.

“OK, let's draw them!” I said.

Most children drew stick figures, but some were considerably more elaborate. Some sisters were lined up across the page, and others were more irregularly placed.

“When you get your sisters drawn, raise your hand and I'll come and take a look,” I said. After I had looked at several, I drew three stick-figure sisters on the board.

“What's next? Who can remind us about what we are trying to find out?” I asked.

“How many mice!” responded several children.

“OK, what shall we do?” I asked. “Do we need to read the story again?”

“Each sister had four cats,” said Allison. She continued, “Maybe we can draw four cats for each sister.”

“Whew, that is a lot of cats,” I replied. “But it is a good idea. If you don't want to draw a whole cat, you can just put a *c* for each one,” I suggested. The children took off here. Stick-figure cats and letter *c*'s sprouted around each sister. Some children just wrote “4 c” for four cats. After I looked at the children's drawings, I drew “c c c c” under each sister on my picture on the board.

“What next? Who can say one more time what we want to know?”

“How many mice!” said several.

“OK, who will read the part of the story about mice?” I asked.

“Each cat caught five mice,” responded several children.

“Oh, no,” said Jonathan. “That is too many mice.”

“How are we going to do it?” I asked.

“We will have to put five mice around each cat,” said Lacey.

“That is a lot,” I said. “If you want, instead of drawing each mouse, you can just put an *m* for each one.”

The *m*'s were rather crowded and disorderly on most drawings. After looking at some pictures, I put mine in straight columns. My board drawing looked something like this (here I will use letter: *s* for sister, *c* for cat, and *m* for mouse):

	s				s				s		
c	c	c	c	c	c	c	c	c	c	c	c
m	m	m	m	m	m	m	m	m	m	m	m
m	m	m	m	m	m	m	m	m	m	m	m
m	m	m	m	m	m	m	m	m	m	m	m
m	m	m	m	m	m	m	m	m	m	m	m
m	m	m	m	m	m	m	m	m	m	m	m

“What do we want to know?” I asked.

“How many mice!” replied the children.

“OK, we have our calculators. How can we find out?” I asked.

Programs that the children thought of:

[5][+][5][+][5][+][5][+][5][+][5][+][5][+][5][+][5][+][5][+][5][+][5][=]

[20][+][20][+][20][=]

[1][+][=][=][=][=] . . . (sixty times)

I asked them, “Can you figure out how to do it on the calculator without using the plus key?'' The children had not been formally introduced to multiplication, but they knew a little about the times key. Still, no one suggested a way to do it.

''How many cats did the sisters have?'' I asked.

''Twelve,'' several children answered.

''OK, put 12 into your calculator. Now, how many mice did each cat catch?''

''Five.''

''Here is a new key for you: It is 'times.' Try [*][5][=].''

The children were excited about this shortcut.

''Can you think of any other way?''

Eventually, with some help, the children figured out [20][*][3][=]. But no one came up with [3][*][4][*][5][=], so at the end I showed them.

In this lesson children learned that there are many ways that one can get the right answer–many different sequences of button presses (programs) can be used. They were encouraged to look for new ways, and they were very interested in trying out ideas. It was their first introduction to programming with the calculator.

When I Went to St. Ives

Once upon a time, in the city of Trento, in Italy, lived three sisters.

Each sister had four cats.

Each cat caught five mice.

How many mice were caught by the sisters' cats in the city of Trento in Italy?

CHAPTER 6

Clay Families

This lesson has been taught in several second-grade classrooms. Present were an instructor and the regular classroom teacher. It took between forty-five and fifty-five minutes. It has been taught in kindergarten and first grade as well.

In this lesson we divide clay into two equal parts, make a ball from one part and divide the rest into two equal parts, make a ball from one part, and continue this process until the pieces of clay are too small to handle.

PROPS

- calculators
- rulers (with centimeter and/or inch scales)
- about 6 oz. of clay per person (we used Play-doh)
- plastic knives for cutting clay (optional; rulers can be used instead)
- large calculator poster visible to all

PART 1. TO TAKE HALF OF SOMETHING, YOU CAN DIVIDE IT BY TWO

We first handed out only the calculators and began with the "first/second" game. The instructor wrote "first" and "second" on the board and explained, "The game works this way. I am going to choose one number as the first and another as the second. I choose 5 for the first, and 10 for the second. Now you need to put 5 on the display of your calculator. So press the 5 key. Now, the game is to change the 5 into the second number, 10. But you are not allowed to press the on-clear key and just enter the number 10. You have to do something to 5 to make it into 10. What can you do?"

Almost all children said you need to add five in order to get 10. So the keystrokes are [+][5][=], and the display shows 10. The instructor wrote on the board $5 + 5 =$, and showed, using the poster, which keys to press. She especially pointed out the plus and equals keys, and named them.

"Here is the next part of the game. Now we want to change 10 back into 5. How can we do it? It is not fair just to press the [ON/C] key and then enter 5. What do you have to do to 10 to change it into 5?"

The kids figured out that you have to take away 5. The instructor wrote on the board, $10 - 5 =$, pointed to the keystrokes on the poster: [−][5][=], and said, "Minus five equals. Try it, and see what you get on your display." The children pressed their keys and said, "Five."

ILLUSTRATION 4. A clay family.

"OK, here is the next part of the game. How can you change 5 back into 10, but this time the rules change and you have to do it without using the plus key? What can you do to 5 to make it 10?"

One or two children said maybe you can times it. They weren't sure, and they experimented with what number to "times it by." Eventually they had the solution: $5 \times 2 = 10$. The instructor wrote on the board $5 \times 2 =$ and showed the keystrokes [*][2][=], saying, "times two equals." She invited all children to try it and see what they got on their display.

"Now, let's change 10 back into 5, but this time it's illegal to use take away. How do you do it without using minus?"

This was hard. One child said you need to take half of 10 to make it 5. But they had not been introduced to division before. "Is there another key on the calculator that will help us? How about the divide key?" The class experimented and figured out that 10 divided by 2 gave 5. The instructor wrote on the board $10 \div 2 =$ and showed the keystrokes: [÷][2][=], saying, "divided by two equals." The teacher said, "So to take half of 10, you can divide it by 2."

"Now let's play the game again. Will someone in the class suggest a new first number?" Emily suggested 16. "How about a second number?" Danny wanted 30, but the teacher said that might be hard and suggested 32. We played the first/second game again. The numbers were more difficult this time, but the children easily figured out two ways to change 16 into 32: by adding 16 or by multiplying by 2. Similarly, they figured out two ways to change 32 back into 16: by subtracting 16 or by dividing by 2. For each, the instructor wrote on the board as she had above, and she named the keystrokes and pointed to them on the poster.

PART 2. MAKING CLAY FAMILIES

Next we handed out the Play-doh (one can per child), plastic knives, and rulers. "Take the clay out and smoosh it around and get it so you can roll it." The children simply played with the clay for a few minutes.

"OK, now roll your clay into a sausage. It doesn't matter how long. Just get it into a nice even sausage.'' The instructor demonstrated by rolling her clay into a cylinder of uniform diameter. Children needed some help with this; pieces of the clay would come off, and they had some trouble making their sausages even. But everyone could do it, sometimes with the teacher's or instructor's help.

''OK, has everyone got a sausage? Now we want to cut it into halves, into two equal pieces. We want to be sure that they are equal. How are we going to do it? We have rulers, and we have calculators.'' This was a challenge. Some children tried to bend their clay, but it did not work well because the sausages were too fat. Some children realized they could measure the lengths of their sausages with their rulers. The instructor reminded them about how you can get half of a number by dividing it by two. ''How can we use our calculators to help?'' The light dawned for several children. They said, ''We can measure the length of the sausage, put the number in the calculator, and divide it by two.''

The task was not so easy. Some children did not know how to measure the lengths of their sausages. They needed help putting the end of the sausage at the zero-end of the ruler. The instructor told them it did not matter which side of the ruler they used (inches or centimeters), but that it was important to put their sausage even with the end of the ruler. Some children needed help here. For example, one child said, ''My sausage ends on the 7. How do I get half?'' The instructor said, ''Put 7 into your calculator and divide it by 2.'' The child did so, and got the answer 3.5.

''What does it mean?'' he asked.

''Well, look at your sausage.'' (It was still next to the ruler.) ''About where do you think the halfway point is?''

''Between 3 and 4.''

''That's what 3.5 is: it is exactly halfway between 3 and 4. So make a mark with your knife or fingernail at 3.5 (where the long black line is on your ruler) and cut your sausage there.''

In a case when a child's sausage did not end exactly on an integer, the teacher or instructor helped the child stretch or mash the sausage so that it more or less did so. The child then entered the integer into the calculator and divided it by 2.

''Does everyone now have two equal pieces? Here's where the fun starts. Roll one of the pieces up into a ball. It will be your daddy ball.'' The instructor demonstrated.

''Put the ball out of your way on your desk. Now take the remaining sausage and cut it into two equal pieces! You can do this just by measuring what you have and cutting it in half, or you can roll it out to be a thinner sausage and measure and cut.'' She demonstrated.

This time the children performed the calculator computations without much difficulty. There was a small problem in the case when a number such as 3.5 was again divided by 2, yielding 1.75. The teacher recommended that children roll their clay out so that its length could be approximated by an integral number of inches (or centimeters). We began to lose some precision here.

''Now, does everyone have two halves again? Now roll one half up into a ball. It will be the mama ball. Put it out by the daddy ball, and cut the remaining sausage into two equal pieces.'' She demonstrated.

After the first two measurements, children began to approximate their divisions and not use the calculator. ''Now, roll one half into a ball. It will be the biggest kid ball. Put it next to the mama ball. Take the remaining piece of clay and divide it into two equal pieces. . . .''

The children just took off with this. They were extremely excited to see their families of balls grow. At one point the classroom was actually quiet as they worked. The instructor had brought to class the family of balls she had made beforehand, and she put them out on a desk for the children to see.

"Stop when your clay gets too small!"

There was teamwork here, and some children with their desks pushed together lined up their families together. The clay came in four colors. At the beginning we had made sure that for children working in groups, each had a different color of clay.

When everyone had finished, the instructor said, "Here is a question for you. Suppose I take all the family from the mama on down (not the daddy), and roll it up into a ball. How big will it be?" The kids knew the answer to this. They said, "It will be as big as the daddy, because that's how we started out!" Then she asked, "Suppose I took all the family from the biggest kid on down (not the daddy and not the mama), and rolled them up into a ball. How big would the ball be?" The kids knew this too. They said it would be as big as the mama, because that's how we cut it.

One child almost shouted, "This is more fun than going to the fair!"

When it was time to quit, the instructor asked the children to put their clay back into the cans. This caused some alarm. "We do not want to smoosh the babies!"

In one class, we added up the number of ball-people that had been made by the children. The teacher went around to each desk and each child named his or her number. We all kept a running total on our calculator. The final total was 287.

REMARKS

(1) Plastic knives are really not needed. Fingernails for marking on clay, and rulers for cutting it, work fine.

(2) Some children wanted the pink (almost fluorescent) clay. Only a quarter of the cans contained pink clay, and this created some problem. But when we use clay again, the instructor will request that those who had pink before now take another color.

(3) The instructor could make only fourteen ball-people before her clay became a tiny dot. Her smallest ball was thus $1/(2^{14})$ or about 1/16,000 as big in volume as the first one. Some children cheated here—they did not follow the instructions, and they got twenty-five or more ball-people. (This is physically impossible, if you follow instructions, since such balls would weigh 1/5,000,000 oz. or less.) We did not want to make it a competition. The idea was not to get a lot of clay people.

CHAPTER 7

Money

This lesson was taught in some first- and second-grade classes. In the lesson children use real coins.

PROPS

- one dollar in change for each child: ten pennies, five nickels, four dimes, and one quarter (These can be put in envelopes and passed out.)
- two paper plates of different colors or sizes for each child (We used red and green salad-sized plates.)
- magnifying glasses for looking closely at the coins (one per child)
- one half-dollar coin, a Susan B. Anthony dollar coin, and one two-dollar bill (You can get these at a bank.)
- calculators
- calculator poster, pinned to a bulletin board visible to all, or overhead calculator

THE LESSON

As the envelopes of change are passed out, the teacher can tell the children that each envelope contains one dollar in change and that each child is checking out an envelope on the honor system. Each child should print his or her name on the envelope he or she receives. At the end of the lesson, the teacher can check that each envelope contains one dollar, and if change is missing, the person whose name is on the envelope can be held responsible for the money.

Topic 1. Some Background about Money

A group discussion about why we have money and what it is for can be held. We asked the children: if you go back in history, did people always have money? What did people do before they had money? The concept of bartering can be introduced.

Coins have to be manufactured. What is the name of the place where coins are made? (It is a mint.)

ILLUSTRATION 5. Some props for the money lesson.

Topic 2. Looking at Coins

PENNIES

Pass out magnifying glasses. Ask children to dump out their coins onto a plate and examine a penny. Hold up a penny in case some children can't pick it out. Ask them to look at the penny with their magnifying glass and describe what they see. Start with the front side. There is a picture of Abraham Lincoln. Over the top of his head are the words, "IN GOD WE TRUST." There is also the word "LIBERTY"; the date indicating the year the coin was minted; and on some coins a tiny letter, D, under the date, indicating that it was made at the Denver mint. Ask children if they can see the letter. (There are two other mints in the United States, in San Francisco and Philadelphia. Pennies minted in San Francisco have an S, and those minted in Philadelphia have no mint mark.)

Now look at the back of the penny. Ask children if they know what the building is. (It is the Lincoln Memorial in Washington, D.C.) On some pennies, using a magnifying glass, one can see the statue of the seated Lincoln between the center two columns. Across the bottom are the words "ONE CENT," and across the top is "UNITED STATES OF AMERICA," and "E PLURIBUS UNUM," Latin for "one from many." Discuss what this means. Also state that a penny is worth one cent.

Find the oldest penny in the classroom. This is a fun thing to do. In one first grade I found that children would say, "I have a 1995 penny! I have the oldest!" In other words, they thought oldest meant the penny with the "highest" number. Explain that lower numbers mean older. We found that the oldest penny in our class had the date 1960. How old is that? If this year is 1995, how can we compute, using calculators, how old the 1960 coin is?

[1995][−][1960][=]

The coin is 35 years old.

Note: it is not obvious to children what one needs to do to calculate how old the coin is. In particular, it is not obvious to them that the operation should be subtraction.

NICKELS

Now examine a nickel. (Again, hold one up.) The front side shows the face of Thomas

Jefferson. (He has a pony tail!) There are the words, "IN GOD WE TRUST" and "LIBERTY." Also, there is the date that the coin was minted and on some coins we found the letter D, indicating that it was made in Denver. On the back is Monticello, the place in Virginia that Jefferson had built. (Its name is printed under the building.) The words, "E PLURIBUS UNUM," "UNITED STATES OF AMERICA," and "FIVE CENTS" also appear. State that a nickel is worth five cents. So a nickel has the same value as five pennies.

DIMES

Next, examine a dime. (Hold one up.) It is the smallest United States coin (in physical size). On the front is a picture of Franklin Delano Roosevelt. Again, we find the words "LIBERTY" and "IN GOD WE TRUST," a date when the coin was minted, and often a mint mark. (We found a D.) On the back are a torch and some laurel and oak leaves. There are the words, "UNITED STATES OF AMERICA," "ONE DIME," and "E PLURIBUS UNUM."

Note that the word penny does not occur on the penny, and the word nickel does not occur on the nickel. But the word dime does occur on the dime. A dime is worth ten cents. It has the same value as ten pennies or two nickels. Even though it is smaller than a nickel, it takes two nickels to equal a dime in value. The oldest trick in the book is for older siblings to trade their nickels for their younger siblings' dimes. Tell children not to fall for it!

QUARTERS

Now, examine a quarter. (Hold one up.) On the front there is a picture of George Washington (he also has a pony tail), the words "LIBERTY" and "IN GOD WE TRUST," a date when minted, and often a mint mark. (We found a D.) On the back are a picture of an eagle with its wings spread, with the words "UNITED STATES OF AMERICA," "E PLURIBUS UNUM," and "QUARTER DOLLAR." A quarter is worth twenty-five cents. It has the same value as twenty-five pennies, five nickels, or two dimes and a nickel.

Although children do not have a half dollar in their collection, it should be mentioned that the largest United States coin in circulation is called a half dollar. If you have one to hold up, that is good. It has a picture of John F. Kennedy on the front, and an eagle with shield and the Presidential seal on the back. It is worth fifty cents, the same as two quarters (or five dimes, ten nickels, or fifty pennies).

You can briefly talk about paper money. Which denominations do we have in the United States? If you have a two-dollar bill, show it! Show also a Susan B. Anthony dollar coin if you have one.

Topic 3. Counting the Money in Each Envelope

Begin using calculators. How much money is in each envelope? Let's start with pennies. How many? Ten. How many cents? Ten cents. (Write on the board: ten pennies are worth ten cents.) How many nickels? Five. How many cents are they worth? Have children suggest ways to calculate this, and then show them several ways:

[5][+][5][+][5][+][5][+][5][=]	Display: 25
[5][+][=][=][=][=][=]	Display: 25
[5][*][5][=]	Display: 25

Five nickels are worth twenty-five cents. (Write on board.)

How many dimes? Four. How many cents are they worth?

[10][+][10][+][10][+][10][=] Display: 40
[10][+][=][=][=][=] Display: 40
[10][*][4][=] Display: 40

Four dimes are worth forty cents. (Write on board.)

One quarter. How many cents is it worth? Twenty-five. One quarter is worth twenty-five cents. (Write on board.)

Now let's determine how much money we have altogether.

[10][+][25][+][40][+][25][=] Display: 100

We have 100 cents. How much is that? One hundred cents is the same as one dollar. (Note: lead the children through the sequence of keystrokes to get the sum 100. We found that children would forget to press the plus-key. Show this clearly using the calculator poster or overhead calculator.)

Topic 4. Putting Amounts of Money on a Plate

"Dump out all your coins onto your red plate. Now, by taking coins from the red plate, can you put forty-four cents on the green plate? Do it! How did you do it? Which coins did you use?"

Keep a record on the board, writing the number of pennies, nickels, dimes, and quarters. Most children will say they have four dimes and four pennies. Write four dimes make forty cents; four pennies make four cents. Have children add the two numbers.

[40][+][4][=] Display: 44

"Can you make forty-four cents without using dimes?"

Have children offer solutions. Keep a record on the board of the coins selected, and have the children add together the amounts from the different coins. One solution: one quarter, three nickels, and four pennies. Write:

one quarter makes twenty-five cents
three nickels make fifteen cents
four pennies make four cents

[25][+][15][+][4][=] Display: 44

Ask for other solutions. Always write on the board, and have children add to see if the total is forty-four. We found that children are quite surprised that different collections of coins can have the same value.

"Dump all your coins back on the red plate. Now, can you put fifty-six cents on the green plate? How?" Again, list the solutions.

"Can you make fifty-six cents without using pennies?" This is a great challenge for the first grade! Many think it is possible and only see that it is not when they actually try.

Topic 5. Figuring the Cost of a Pack of Gum

In the grocery store there was an ad for three packs of gum for one dollar. How much would one pack of gum cost? In classes where this was tried, no one tried the keystrokes

[100][÷][3][=]. Instead, they guessed. The first guess in one class was eighty cents. If one pack costs eighty cents, how much will three packs cost?

[80][+][80][+][80][=] Display: 240

Three packs would cost 240 cents, or $2.40. That is too much! Children guessed until they knew that at thirty-three cents a pack, three will cost ninety-nine cents, and at thirty-four cents a pack three will cost $1.02.

We then used a calculator:

[100][÷][3][=] Display: 33.333333

Each pack of gum will cost a little over thirty-three cents (thirty-three and one-third, to be exact). So the store will probably charge you thirty-four cents for the first pack and thirty-three cents for the second and third packs. Let's add:

[34][+][33][+][33][=] Display: 100

Ask children to put their coins back in the envelopes. Have them check (and write on the board) that they have ten pennies, five nickels, four dimes, and one quarter. Collect envelopes and plates.

CHAPTER 8

Calculator Memory

THE LESSON

Children are going to perform addition and subtraction mentally and check the result with the calculator, using the memory key, [M+]. The problems should require one or more operations and can be either abstract (subtract 3 from 11) or concrete. (You had fifty cents. You spent ten cents. How much do you have left?)

Only the numbers should be written on the blackboard, not the story. It is up to the teacher to decide whether children are allowed to use manipulatives (or fingers) as a help in mental computations.

The procedure shown below by examples should be carefully explained and illustrated.

Example 1

The teacher says, "Subtract three from eleven" and writes 11 and 3 on the blackboard.

(1) Children compute the difference on the calculator without seeing the result.

Keystrokes:	Comments:
[MRC][MRC]	Be sure that memory is cleared.
[11][M+] [3][M−]	The calculator has already subtracted the numbers, but the result is hidden.

(2) Children mentally compute 11 − 3 and write down the answer.

(3) Children compare their answer to the calculator's answer by pressing [MRC] (or [MRC][MRC]) and reading the display.

Example 2

The teacher says, "You had fifty cents and you spent ten cents. How much do you have left?" and writes 50 and 10 on the blackboard.

(1) Children compute the difference on the calculator without seeing the result. Keystrokes: [MRC][MRC] [50][M+] [10][M−].

(2) Children mentally compute 50 − 10 and write down the answer.

(3) Children compare their answer to the calculator's by pressing [MRC] and reading the display.

Example 3

The teacher says, "Add the numbers" and writes 1000, 100, 10, and 1 on the blackboard.

(1) Children compute the sum using the keystrokes: [MRC][MRC] [1000][M+] [100][M+] [10][M+] [1][M+].
(2) Children mentally compute 1000 + 100 + 10 + 1 and write down the answer.
(3) Children compare their answer to the calculator's by pressing [MRC] and reading the display.

Example 4

The teacher says, "Subtract ten from two" and writes 2 and 10 on the blackboard. The steps are the same as above, but special attention should be paid to:

- Keystrokes: children must press [2][M+] [10][M−] and not [10][M+][2][M−].
- Reading the display: the minus sign is hidden under the letter M. It is better to read the display by pressing [MRC][MRC] and exposing the minus sign.

REMARKS

(1) The examples given here are arbitrary. The teacher should determine the difficulty and content of the problems.
(2) After the children learn the procedure, the problems can be given in rapid succession.
(3) Children can work in pairs, one doing mental computation and the other handling the calculator. The sequence of keys remains the same.
(4) One explanation of how the memory key works goes as follows: there is a place inside the calculator where it can write down a number without showing it. This place is called memory. If you want to see what is there, press [MRC]. [M+] adds the displayed number to the number already in the memory. [M−] subtracts the displayed number. Having zero in the memory means that the memory is empty. Pressing [MRC][MRC] not only shows what is in the memory but also makes it empty.

CHAPTER 9

Number Game with Alphabet

This lesson introduces addition and comparison of negative numbers. It was designed and taught in fourth grade by Marilyn Sprinkle on the basis of an idea suggested by Penny Vogel.

(1) Assign an integer to each letter of the alphabet. Consonants get positive values (or 0), and vowels get negative values. Example:

a	b	c	d	e	f	g	h	i	j
−2	1	2	3	−1	4	5	1	−3	2
k	l	m	n	o	p	q	r	s	t
3	4	5	1	−4	2	3	4	5	1
u	v	w	x	y	z				
−5	2	3	4	5	1				

(2) To get the value of a word, simply add the values of its letters. Example: look at the word "alphabet":

a	l	p	h	a	b	e	t
−2	4	2	1	−2	1	−1	1

$-2 + 4 + 2 + 1 + -2 + 1 + -1 + 1 = 4$

The value of the word "alphabet" is 4.

EXAMPLES OF WORDS AND THEIR VALUES

Word	Value
advice	$-2 + 3 + 2 + -3 + 2 + -1 = 1$
notice	$1 + -4 + 1 + -3 + 2 + -1 = -4$
police	$2 + -4 + 4 + -3 + 2 + -1 = 0$
remain	4
fountain	−7
alphabet	4

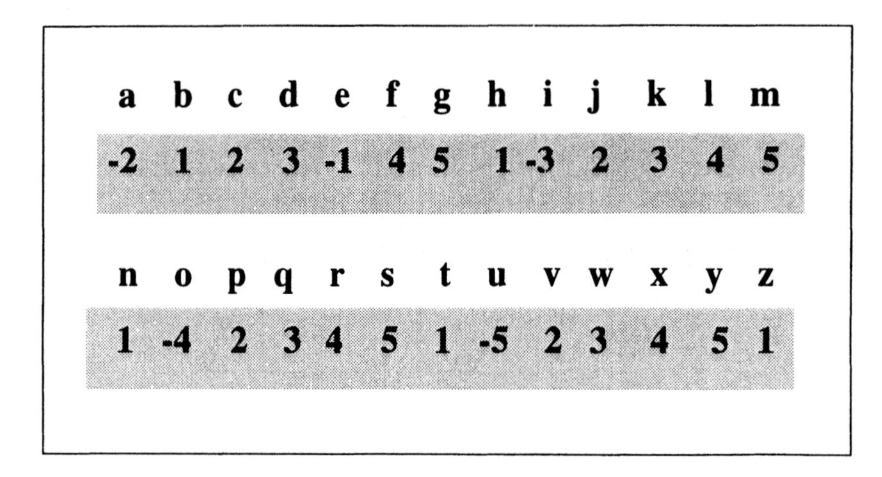

ILLUSTRATION 6. Assigning an integer to each letter of the alphabet.

EXAMPLES OF QUESTIONS AND PROBLEMS

Find a word having a value of 3.

Which word above has the highest value? ("remain" and "alphabet," each with a value of 4)

Can you find a word with a higher value?

Which word above has the lowest value? ("fountain," −7)

Can you find a word with a lower value?

What is the value of your first name? Your last name?

COMPUTATION

Computation should be done mentally or with calculators. This lesson is not an exercise in written calculations. When using calculators in this lesson, the minus key should NOT be used! The minus key is for subtraction. The plus-minus key gives a minus sign to negative numbers.

First Method

Enter negative numbers using the [+/−] key, and then add using the [+] key.

Example: "advice," [2][+/−][+][3][+][2][+][3][+/−][+][2][+][1][+/−][=]

Second Method

Use [M+] and [M−].

Example: "notice," [1][M+][4][M−][1][M+][3][M−][2][M+][1][M−][MRC]

A combination of mental calculations with some use of the calculator is probably the best method for computing the numerical value of a word.

REMARKS

(1) *Comparison of negative numbers:* We suggest that $-7 < -2$ be spoken as, "Negative seven is lower than negative two" or "Minus seven is lower than minus two." Using the terms "bigger" and "smaller," which are more standard, can be confusing. Even mathematicians often call -1000 a "big negative number" and -1 a "small negative number." So saying that a big negative number is smaller than a small one is at least strange. Also the expression "small difference" means that the difference is close to zero and not that it is a "big negative number." The terms "high" and "low" do not carry any connotation of size and are in agreement with everyday references to temperatures.

(2) *History:* Assigning numbers to letters of an alphabet has a long history. The ancient Greeks and other people living around the Mediterranean Sea did not use special symbols to write numbers. They wrote numbers using the letters of their alphabet. The first nine letters stood for the numbers 1 to 9: a = 1, b = 2, c = 3, d = 4, e = 5, f = 6, g = 7, h = 8, and i = 9. The next nine letters meant tens: j = 10, k = 20, l = 30, etc. The next group stood for hundreds: s = 100, t = 200, etc. Only after all letters were used were other methods for writing numbers applied. So, for example, the words jet and tej both meant 215 = 200 + 10 + 5.

Of course, the Greeks used the Greek alphabet, and the Hebrews used the Hebrew alphabet. Neither used the Roman alphabet, which we now use. One result of these systems of writing numbers was that all words have numerical values such as presented in the lesson above. Some people attached magical meanings to these numbers, as astrologers attach to the dates of birth. So, depending on the number attached to your name you could be lucky or unlucky, and so on. If you want to know more about one very famous magic number, 666, called "the number of the beast," you may read "The Number of the Beast" in *Puzzles from the Other World,* by Martin Gardner, Vintage Books, 1984.

CHAPTER 10

Introduction to Measurement (First Grade)

PROPS AND TOOLS

You will need rulers with centimeters and millimeters, pencils, blank sheets of paper – one for each child. Children will also need crayons or colored markers.

THE LESSON

(1) Looking at rulers. How long is one centimeter? How long is one millimeter? Five millimeters make one half of a centimeter, ten millimeters make one centimeter. When you count centimeters and millimeters, you count spaces between lines, not the lines.

(2) Terminology. (Teacher writes on the blackboard; children just read and speak.)

cm is read centimeter; mm is read millimeter.
2.3 cm is read two centimeters and three millimeters.
6 mm is read six millimeters.
etc.

Children should read aloud 6 mm, 3.4 cm, and so on and show the length on their rulers. They do not have to show it starting at the zero point of their rulers; 2.5 cm can be shown between the 13 and 15.5 cm markers on a ruler. This activity can be carried out until children lose interest.

(3) Children work in pairs. Each pair marks three points on a blank sheet of paper (not too far apart) and draws a triangle, using rulers to make straight lines. The children should work cooperatively. It is easier to draw a line if someone helps you hold your ruler. The word "triangle" should be used, and also the words "sides of a triangle," and "vertices of a triangle."

(4) Both children in each pair measure each side of their triangle, one side at a time, using rulers and counting centimeters and millimeters. They do not have to read the numbers on the scale, but they may. They should reach agreement about the lengths.

REMARKS

Children may need help with tasks 3 and 4. Tasks 3 and 4 should be repeated until children get proficient or lose interest. Talking should be encouraged. The teacher may help by drawing on the blackboard triangles of different shapes and sizes in different positions. "Now I'll draw a tall skinny triangle upside down."

Each pair of children should fill two sheets of paper. The number of triangles on each sheet is not important.

The lesson should end with each pair coloring the triangles on their sheets. Each child will then have one sheet with colored triangles to take home.

There is no hidden agenda in this lesson. Children are learning to draw triangles, measure their sides, and talk about what they are doing using acceptable terminology.

CHAPTER 11

Menu

This lesson has been taught in several first- and second-grade classes.

In this lesson children will read a menu and calculate prices of different combinations of menu items.

PROPS

- a calculator for each child
- an overhead calculator for the teacher
- a menu (see end of chapter) prepared as a transparency and as a handout for each child

THE LESSON

"First, let's read the menu. What's the name of the restaurant? Who can read the breakfast main dishes? Lunch and dinner main dishes? Sides? (What is a side?) What is a bagel? What is coleslaw? Who can read the drinks? And what's that last item? Who can read the desserts? What's the cheapest item on the menu? (A bagel; it costs forty-five cents.) What's the most expensive item on the menu? (Bacon and eggs; they cost $2.59.) Can you find two items that cost the same amount? (Grilled cheese and fish sandwich each cost $1.50. Hot chocolate and apple pie each cost seventy-nine cents.)

"Now, let's use our calculators to compute what some foods will cost. How much does a hamburger and cola (coke) cost? First, the hamburger costs $1.75. How do we enter $1.75 into the calculator? We have one dollar and seventy-five cents. A decimal point separates the number of dollars from the number of cents. You use the decimal point key on the calculator to separate the number of dollars from the number of cents. So you enter [1][.][7][5]. Now we want to add the price of a coke: sixty-two cents. So we first enter plus: [+]. How do we enter sixty-two cents? We must use a decimal point: [.][6][2], and finally we must press the equals key, [=] (Display: 2.37). A hamburger and coke cost $2.37."

Note: explain what happens when you don't use a decimal point in entering 62. It is treated as sixty-two dollars!

Now try various two-, three-, and four-item meals, such as pancakes, orange juice, and milk; cheeseburger, french fries, and a shake; etc. Here, you will need to explain that, for example, .7 on the display is interpreted as seventy cents; 3.9 is interpreted as three dollars and ninety cents, etc. You might also have the children pick a meal for themselves and calculate how much it costs.

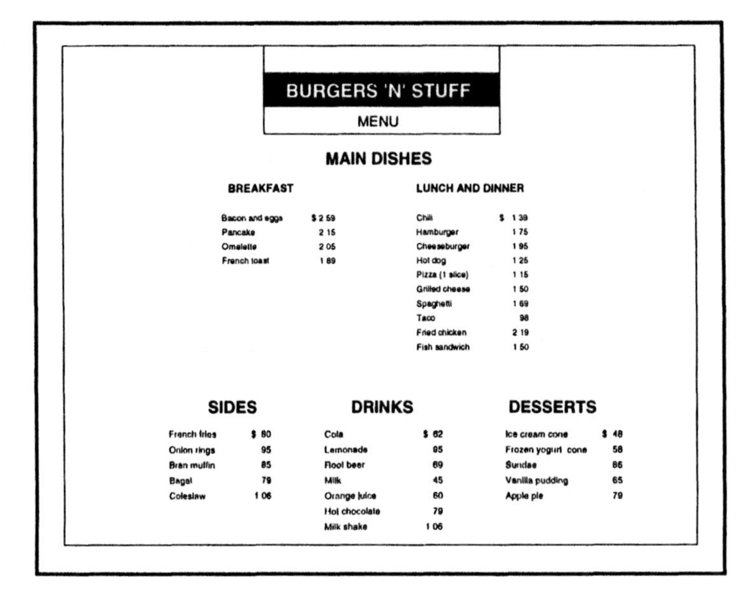

BURGERS 'N' STUFF

MENU

MAIN DISHES

BREAKFAST	
Bacon and eggs	$ 2 59
Pancake	2 15
Omelette	2 05
French toast	1 89

LUNCH AND DINNER	
Chili	$ 1 39
Hamburger	1 75
Cheeseburger	1 95
Hot dog	1 25
Pizza (1 slice)	1 15
Grilled cheese	1 50
Spaghetti	1 69
Taco	98
Fried chicken	2 19
Fish sandwich	1 50

SIDES	
French fries	$ 80
Onion rings	95
Bran muffin	85
Bagel	79
Coleslaw	1 06

DRINKS	
Cola	$ 62
Lemonade	95
Root beer	69
Milk	45
Orange juice	60
Hot chocolate	79
Milk shake	1 06

DESSERTS	
Ice cream cone	$ 48
Frozen yogurt cone	58
Sundae	86
Vanilla pudding	65
Apple pie	79

ILLUSTRATION 7. Burgers 'n' Stuff menu.

Problems

(1) Suppose a big football player orders four hamburgers. How much will he pay? Most children will try

[1.75][+][1.75][+][1.75][+][1.75][=] Display: 7.

He will pay seven dollars.

Show them how to use multiplication:

[1.75][*][4][=] Display: 7.

Explain that 1.75 times 4 is the same as 1.75 plus 1.75 plus 1.75 plus 1.75.

How much would the football player pay for 3 root beers?

(2) The teacher wants to treat everyone in the class to an ice cream cone. How much will she pay? Use the actual number of people in the class.

(3) Five kids each order a taco and a root beer. How much is the total bill?

A solution:

[.98][+][.69][*][5][=] Display: 8.35

The bill is eight dollars and thirty-five cents.

Another solution:

[.98][+][.69][=][*][5][=]

Another solution:

[.98][*][5][M+][.69][*][5][M+][MRC]

REMARKS

Here are some things we learned from teaching the lesson:

(1) Having a large calculator poster in the front of the room is a big help, especially in explaining the decimal point, times key, M+, and MRC.

(2) Children like to figure out how much their own choices cost.

(3) Children make lots of errors in entering numbers. Often they put something in memory and "can't get it out." It is a good thing to teach them

[MRC][MRC][ON/C][ON/C]

as a way to completely clear out their calculator, when things go awry.

(4) Children will need help in interpreting dollars and cents from the calculator display. They often forget the decimal point for items that cost less than a dollar, and then of course the numbers are huge.

(5) Suppose we wanted to know how much one of each of the breakfast items cost. Here you have a choice of using [M+] between prices of items and [MRC] at the end, or [+] between prices of items and [=] at the end.

(6) Bring in real menus from restaurants and fast food places in your area. Children will take off with this.

(7) If there is a tax on food in your city or state, you can easily add it with a calculator. Suppose the bill (before tax) is $8.35, and the tax is 6%. Here are the keystrokes for the total price:

[8.35][+][6][%]   Display: 8.851

The bill with tax is $8.85.

BURGERS 'N' STUFF

MENU

MAIN DISHES

BREAKFAST

Bacon and eggs	$ 2.59
Pancake	2.15
Omelette	2.05
French toast	1.89

LUNCH AND DINNER

Chili	$ 1.39
Hamburger	1.75
Cheeseburger	1.95
Hot dog	1.25
Pizza (1 slice)	1.15
Grilled cheese	1.50
Spaghetti	1.69
Taco	.98
Fried chicken	2.19
Fish sandwich	1.50

SIDES

French fries	$ .80
Onion rings	.95
Bran muffin	.85
Bagel	.79
Coleslaw	1.06

DRINKS

Cola	$ .80
Lemonade	.95
Root beer	1.10
Milk	.45
Orange juice	.60
Hot chocolate	.79
Milk shake	1.06

DESSERTS

Ice cream cone	$ .48
Frozen yogurt cone	.58
Sundae	.86
Vanilla pudding	.65
Apple pie	.79

CHAPTER 12

Shop 'n' Save: Learning to Read a Grocery Bill

This lesson has been taught in several second-grade classes.

PROPS

- a grocery receipt (see end of chapter), pencil, and calculator for each child
- an overhead calculator (optional)
- a transparency of the grocery receipt to use with the overhead projector for the teacher

THE LESSON

"I wonder if you can help me. Last night on my way home from work I stopped at the market and got a few groceries. Here is the bill I got. Can you help me read it and tell me if it is figured correctly?" These remarks generated some curiosity. What store was it? What's all the extra stuff on the receipt?

"Maybe we'd better read it line-by-line and see if we can figure out what it means. Who will read the first line?" Jenny read: "Shop 'n' Save—443-9622."

"What does it mean?" The kids said it was the name and phone number of the store.

"Who will read the second line?" Brad read: "Oh five 29 95 5:25 PM store 61."

"What does it mean?"

The children knew the first part was the date.

"What day is today?"

"May 30."

"What does 05 mean?"

"May."

"Why?"

"May is the fifth month: January, February, March, April, May—1, 2, 3, 4, 5."

"What does 29 mean?"

"The 29th day of May, which was yesterday."

"What does 95 mean?"

"It is the year: 1995."

"What is store 61?"

Kids were stumped here. "Maybe it is the address?" I told them I thought that Shop 'n' Save had a lot of stores, and they were numbered. The one I visited had the number 61.

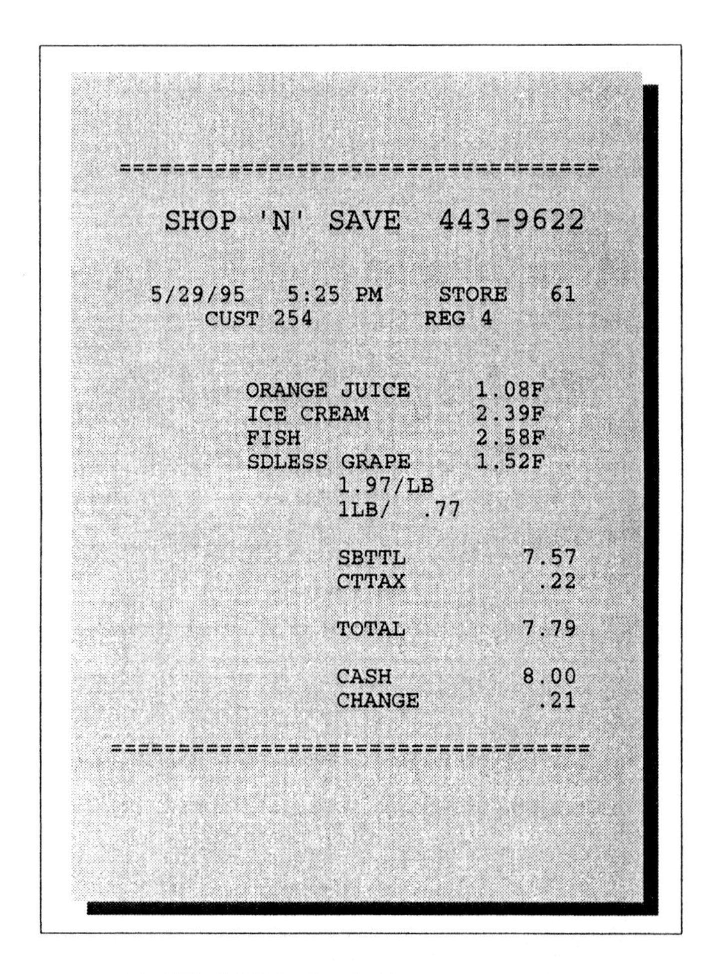

==================================

SHOP 'N' SAVE 443-9622

5/29/95 5:25 PM STORE 61
CUST 254 REG 4

ORANGE JUICE	1.08F
ICE CREAM	2.39F
FISH	2.58F
SDLESS GRAPE	1.52F
1.97/LB	
1LB/ .77	
SBTTL	7.57
CTTAX	.22
TOTAL	7.79
CASH	8.00
CHANGE	.21

==================================

ILLUSTRATION 8. Shop 'n' Save grocery receipt.

''Now let's read the third line.'' Sarah read, ''Cust 254 – Reg 4.''

''What do they mean?'' The children couldn't answer. ''What is a person who goes into a market and buys things called?''

''A customer.''

''So what would it mean, customer 254?''

Danny said, ''Maybe it means you were the 254th customer.''

''Right! Now, how about Reg 4?'' Again, silence. ''When I pay at a supermarket, a clerk takes my money. Where does he or she put it?''

''In a cash register.''

''So what do you think Reg means?'' The light dawned: There must have been at least four cash registers in the store, and I paid at register number four.

''Who will read the next line?''

Angela read, ''Orange juice – 1.08 F.''

''What does it mean?'' The children could read 1.08 as a dollar and eight cents. But they did not know what the F meant. ''I think it means food,'' I said.

Suellen read the next line: ''Ice cream – 2.39 F.'' She read it as two dollars and thirty-nine cents, and the ''F'' she read as food. Fish was next, and it presented no problem for Geri.

Lucas wanted to try ''Sdless grape – 1.52.'' He stumbled over the ''sdless.'' We had a small discussion about grapes, seedless and with seeds, and whether it would be possible to buy just one grape. (This brought some chuckles!)

"Who can read the next two lines?" They presented some difficulty for Bobby. He read 1.97 as one point nine seven, but he did not know what to do with LB.

"In the market, what do we do to grapes to determine how much we pay for them?"

"We weigh them."

"And what units do we use when we weigh them?"

"Pounds."

"Do you know how 'pounds' is abbreviated?"

"LB."

"So how many pounds of grapes do you think I bought?" This was not so easy.

George finally said, "Nearly two pounds."

Lucas read: 1LB/ .77. He did not know what it meant.

"Who can help Lucas? What does 1LB/ .77 mean?"

Again, George said, "One pound must cost seventy-seven cents."

"Okay, if one pound costs seventy-seven cents, and I bought 1.97 lb, then how much should I pay? Does anyone know what we should do on our calculators?" No one did. "Try this on your calculator." I wrote on the board:

[.77][*][1.97][=]. The display shows: 1.5169.

"Who can read it, and what does it mean? How much should I pay?" This brought about some discussion. The children knew I was charged a dollar and fifty-two cents. And they knew that I should have paid 1.5169. The teacher interjected here, "Well, what is 1.5169? Is it a dollar and fifty-one cents and a little more?"

"Yes."

"How much more? Should she be charged a dollar fifty-one, or a dollar fifty-two? Look at the six. What have we learned about rounding?" Some children were able to say it was fair for me to be charged a dollar and fifty-two cents, since six "is bigger than five, so we round up."

"Now Jimmy, can you read the next line?" Jimmy did not know what "sbttl" meant. I told him it meant subtotal.

"Subtotal $7.57."

"How do we know if the subtotal is correct?"

Again, George said, "We add up all the prices."

"Which prices? Tell me, and I will write them on the board, and you can check them with your calculator."

[1.08][+][2.39][+][2.58][+][1.52][=]

We waited until all children were able to get the sum, 7.57, on the display. This took quite a while. Keystroke errors were common, and when a child made a mistake, he or she wanted to start completely over. (The children had not yet mastered how to correct errors on the calculator.)

"Who can read the next two lines?" Amanda read them, but she needed help with "CTTAX" (city tax). "How do we know if the total is correct?" (We add twenty-two cents to $7.57. [+][.22][=] Display: 7.79. It is correct.)

Finally, "Who can read the last two lines?"

Bobby did, "Cash eight dollars. Change twenty-one cents."

"What does it mean?"

"You gave the clerk eight dollars, and you got twenty-one cents change."
"Was my change correct? How do I know?"

[8.00][−][7.79][=] Display: .21 (The change was correct.)

"So, was my grocery bill figured correctly?"
"Yes."
"Of course, maybe the scales didn't weigh correctly, but there is no way we can know that! Next time someone from your family brings home a bag of groceries, get the receipt out and check it!"
We collected the calculators, but the children wanted to keep the receipts.

REMARKS

What to do differently? This was a successful lesson and generated a lot of interest. I would have preferred to have a receipt from a store in the children's town. For a more advanced class, I would have figured out what percentage the tax was.

```
====================================

   SHOP 'N' SAVE  443-9622

 05/29/95   5:25 PM    STORE   61
     CUST 254          REG 4

           ORANGE JUICE     1.08F
           ICE CREAM        2.39F
           FISH             2.58F
           SDLESS GRAPE     1.52F
                 1.97/LB
                 1LB/  .77

                 SBTTL          7.57
                 CTTAX           .22

                 TOTAL          7.79

                 CASH           8.00
                 CHANGE          .21

====================================
```

CHAPTER 13

Comparing Prices

For many canned, bottled, and packaged products, as well as single produce items, both the amount and the price may vary, so a comparison of unit prices is the main method that helps a shopper make reasonable decisions.

EXAMPLES

(1) Tuna is sold in cans of two sizes, a 4-oz can for $1.25, and a 5-oz can for $1.50. Which is cheaper? How much cheaper? How much can I save? Compute and compare unit prices:

Keystrokes:	Display:
[1.50][÷][5][M+]	0.3

The price of the 5-oz can is 30¢ per ounce. Notice that we put this value in the calculator's memory.

[1.25][÷][4][M−]	0.3125

The price of the 4-oz can is a little bit more than thirty-one cents per ounce. The 5-oz cans are a better bargain. The calculator's memory contains the difference of the unit prices.

[MRC]	−0.0125

The 5-oz cans are cheaper by one cent per oz. The minus sign indicates that the first price was smaller. Now if I buy 20 oz (four bigger cans or five smaller cans) how much can I save?

[*][20]	0.25

A quarter.

The shopper can now make a reasonable decision. A reasonable decision does not necessarily mean buying the cheaper product. The shopper may think, "A quarter is not much. I use a can of tuna for each lunch. Five cans will last five days, and four cans only four days. I will buy five smaller cans."

ILLUSTRATION 9. Prices of two sizes of lemon juice.

(2) A 15-lb bag of potatoes costs $3.89, or you pay thirty-four cents per pound when you buy them in bulk. Which is cheaper? How much cheaper? How much can I save?

Keystrokes:	Display:
[3.89][÷][15][=]	0.2593333

Twenty-six cents per pound. Notice that we calculate the cost per pound. We do not buy potatoes by the ounce! Also, because the second unit price is given, we do not need to use the calculator's memory.

[−][.34][=]	−0.0806667

The bag of potatoes is eight cents per pound cheaper. If I buy thirty pounds I save:

[*][30][=]	−2.420001

Two dollars and forty-two cents!

Again, buying more for a cheaper price is not always profitable. The shopper may think: "But I do not need even fifteen pounds. The most I may need is five pounds. 5 × .34 = 1.70. 3.89 − 1.70 = 2.19. So I would pay over two dollars for something I do not need. No, thank you!"

(3) Here is another example from a supermarket of two sizes of the same product for two different prices. Which is a better buy: 20 oz of lemon juice for $1.89, or 32 oz of lemon juice for $2.49?
We want to calculate the price per ounce in each case.

Keystrokes:	Display:
[1.89][÷][20][=][M+]	0.0945
[2.49][÷][32][=]	0.0778125

When you buy twenty ounces at $1.89, you pay about nine and a half cents per ounce. When you buy thirty-two ounces at $2.49, you pay about seven and three-fourths cents per ounce. So the better buy is the 32-oz size. The actual amount saved per ounce, when you buy the larger size, is

[M−][MRC] Display: 0.0166875

This is about one and two-thirds cents.

You could do this problem by estimating. Try not to use your calculator! The price of twenty ounces of lemon juice can be rounded to $2. That gives an approximate cost of ten cents per ounce: $2 \div 20 = 0.1$. (Calculator not needed!) Thirty-two ounces of lemon juice for $2.49 can be rounded to 30 ounces for $2.40. This gives an approximate cost of eight cents an ounce: $2.4 \div 30 = 0.08$. So using these estimates, the amount saved per ounce is two cents.

REMARKS

Have children collect real prices for two (or more) sizes of the same item, and calculate which is a better (or best) buy. Also you can discuss when it makes sense to buy a smaller size of a product, even if it costs more per ounce.

Comparing unit prices and other mathematical techniques can help the shopper make decisions, but they will not tell the shopper what decision he or she should make.

CHAPTER 14

Fractions with Pizza and Money

Each child makes a set of four pizzas (as shown in Illustration 10) and writes on each piece its price.

PROPS AND TOOLS

- for each child, four sheets of paper on which circles of the same size, approximately seven inches in diameter, are drawn (see end of chapter). We used four different colors of sheets and handed them out paper-clipped together.
- scissors for cutting out the circles
- pencils, rulers, and calculators

THE LESSON

We have the following problem: groups of children are buying pizzas, sharing the cost equally. If one pizza costs $3.96, how much should each child in the group pay? Group sizes are three, four, five, or six children.

Tasks

For each task a child has to cut one circle (symbolizing the pizza) into the proper number of equal sections. Children should mark the center of the circle, draw straight lines from the center to the sides, and cut along the lines. Sections should be as equal as possible. It should be stressed that the sharing must be fair. No extra aids such as compasses or protractors should be used. Children should offer their ideas about how to divide the circles. Folding the circle, creasing the folds, and cutting along them is one possibility. After cutting a circle into equal wedges, the corresponding fraction of the price should be computed. Finally, the price of each piece should be written on the piece.

GROUP SIZE 3

Program:

[3.96][÷][3][=]   Display: 1.32

On each of the three pieces, the children should write: one-third of a pizza costs $1.32.

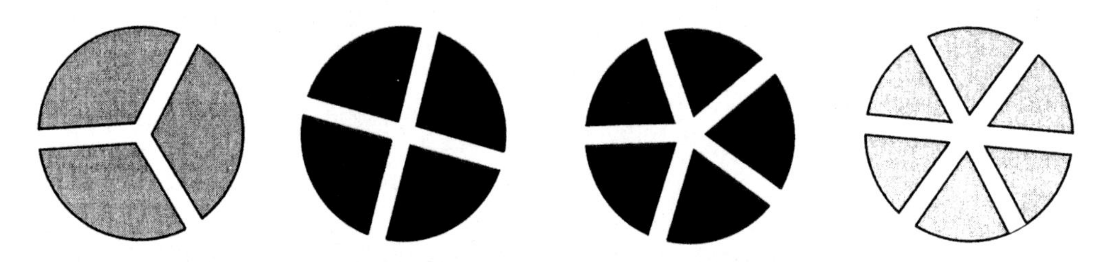

ILLUSTRATION 10. Pizzas divided into thirds, fourths, fifths, and sixths.

GROUP SIZE 4

Program:

[3.96][÷][4][=] Display: 0.99

On each of the four pieces, children should write: one-quarter of a pizza costs ninety-nine cents.

GROUP SIZE 5

Program:

[3.96][÷][5][=] Display: 0.792

On each of the five pieces, children should write: one-fifth of a pizza costs seventy-nine cents.

Somebody has to pay one cent more (eighty cents), because five times seventy-nine cents gives only $3.95.

Program 1:

[5][*][0.79][=] Display: 3.95; one cent is missing.

[+][0.01][=] Display: 3.96; now we have the full price.

Program 2:

[0.79][+][=][=][=][=][=] Display: 3.95; one cent is missing.

[+][0.01][=] Display: 3.96; now we have the full price.

GROUP SIZE 6

Program:

[3.96][÷][6][=] Display: 0.66

On each of the six pieces, children should write: one-sixth of a pizza costs sixty-six cents.

REMARKS

Checking division by multiplication or addition, for example:

[1.32][*][3][=] Display: 3.96

or

[1.32][+][=][=]　　　　Display: 3.96

should be done only if children are not sure that the division gives the correct price for a slice.

At the end of the lesson, each child will have a nice collection of wedges of different sizes and colors. They can paper-clip them together.

The main challenge in this lesson is dividing the circles accurately into equal wedges. Four are easy, but three, five, and six are tricky.

CHAPTER 15

Rectangles

This lesson has been taught in several second-grade classrooms. It takes about forty-five to fifty-five minutes.

Children draw rectangles on one-inch grid paper, cut them out, and determine (using a one-inch square cutout) by counting, and sometimes by grouping and adding, the area of each rectangle in square inches. They then confirm their answers by using a calculator.

PROPS

- calculators
- scissors (one pair per child)
- rulers (one per child)
- pencils
- two or three sheets of one-inch grid paper (see Appendix) (Our paper was hot pink.)
- a large number of square cutouts, one inch on a side, cut from paper colored to contrast with the grid paper (We used white.)
- calculator poster in view of all children

THE LESSON

We prepared beforehand a collection of nine rectangles cut from hot pink paper. They were the following sizes: 1 in. by 2 in., 1 by 3, 1 by 5, 2 by 3, 2 by 4, 2 by 5, 3 by 3, 3 by 4, 3 by 5. We taped them to two sheets of white paper, as shown in Illustration 11, and posted them on the blackboard with magnets. We passed out all the materials except the white square inches.

Part 1. Drawing and Cutting Out Several Rectangles

"Today we are going to start by drawing some rectangles. Does anybody know what a rectangle is?" Children made a sort of rectangle shape with their fingers. One child said, "It is a shape." Another said, "It is like a square, only squashed." Another said, "It has four corners."

The instructor said, "So a rectangle is shaped like a window." The children agreed. The instructor pointed to the picture. "Here are some rectangles. See how I have sorted them:

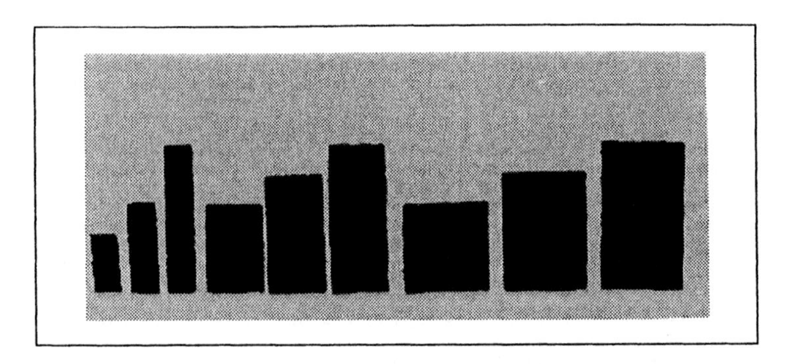

ILLUSTRATION 11. A collection of rectangles.

here is a skinny bunch, then a medium-sized bunch, and then a fat bunch. Each rectangle has a bottom, a top, and two sides. Now you take your rulers and pencils and draw some rectangles on your pink paper. Draw a bunch of different sizes. There is one rule. Each rectangle must have its four corners on dots. And each needs to have a bottom, a top, and two sides.'' Using a piece of pink grid paper that was propped onto the blackboard, the instructor demonstrated how to draw a rectangle that is five inches high and three inches wide. She counted steps as she pointed to and darkened the dots on her paper. ''I start at a corner, and I count one step, two steps, three steps. Then I count down, one step, two steps, . . .'' Using a ruler, she drew lines through the darkened dots.

The children began to draw. Many drew free-hand and did not even try to use a ruler. Their lines were not very straight. They drew quite a few different sizes.

''Now cut them out.'' The children did so, but it was not an easy exercise. They did not cut very straight, and some of the scissors were very stiff and not usable and had to be exchanged for different ones. The instructor waited until every child had at least two rectangles of different sizes cut out.

''Can you figure out, using your ruler, how long your rectangle's bottom is? How tall it is?''

Several children had some difficulty determining how tall and how wide their rectangles were, because they would put a corner on the ''1'' on the ruler, and then count one too many inches. The instructor explained that to do it correctly, you need to count the number of steps you would take (as she had demonstrated earlier). So you line up the end of the ruler with the corner, and then you count steps (inches).

Part 2. Finding the Areas of the Rectangles

''What's the smallest rectangle you can make? Remember the rule that every corner has to be on a dot.'' The children knew, and many had already cut out a rectangle that was one inch square. ''Measure its sides.'' ''Each side is one inch.'' Instructor: ''That's right. It's square, and each side is one inch. We call it one square inch.'' The instructor then taped onto the grid paper on the blackboard a piece of white paper one inch square, so that each corner was on a dot. ''See how it just fits.'' And she gave to each child a one square-inch cutout, from white paper. ''Now we want to give a name to each rectangle that you have cut out. The name we give it is its area. It will be a number—the number of square inches that will fit into it. Let me show you how to count.'' She demonstrated with the three-inch by five-inch rectangle on the board. She moved the white square inch along the interior of the rectangle as she counted: ''One square inch, two, three, . . . fifteen. Now I am going to write the rectangle's name on it: 15. Its area is fifteen square inches." She wrote 15 with a dark pen. "Now you name your rectangles!"

This was a successful exercise–having the white square inches to count with was crucial. Children counted very methodically and carefully. They held the square inch in their fingers and moved it along the rectangle's interior. The teacher and instructor went from desk to desk to help the children. One problem cropped up: the dot paper contained eighty-eight dots (eight dots by eleven dots). To follow the rule that each corner had to be on a dot, a strip on each side of the paper needed to be cut off. Some children had not done this in their original cutting, so the square inch wouldn't fit in the border strip(s). The teacher and instructor had to help here.

Children were shown how to group and add. For a three-inch by four-inch rectangle, they could count $4 + 4 + 4$. For a seven-inch by ten-inch one they could count 10, 20, 30, etc. It was particularly exciting for them to count by tens–they liked the shortcut.

Eventually, children had written a name on each of their rectangles.

Part 3. Computing Areas Using Ruler and Calculator

"Now we are going to see if we gave our rectangles the right names. I named this rectangle 15. Is that right? Is its area fifteen square inches? Let me show you how to tell, using a ruler and calculator. Let's measure its bottom and its side. Three inches and five inches. What can we check on our calculator to see if its name should be 15?"

The children suggested, and the instructor wrote on the board: $3 + 3 + 3 + 3 + 3 =$. She showed the key presses on the calculator poster. The children were excited to see 15. "Is there another way to do it?" The children suggested $5 + 5 + 5$, and the instructor wrote it on the board and showed again the key strokes on the poster. "Now, can you use the calculator one more time, but without using the plus key?"

Some children suggested, "You can times it."

"Times what?"

"Three times five." The instructor wrote $3 \times 5 =$ on the board, and showed the keystrokes on the poster. "Now, you check your names. Measure the bottom and side of your rectangle, and use your calculator." Children were able to follow this, and they did it with great enthusiasm.

Some children who had made the biggest possible rectangle knew that it had eleven dots on one side and eight on another. A few thought 11×8 would give the area in square inches, and they said they got eighty-eight and not seventy, which they had counted before. The teacher helped them correct their error, by placing the corner of the rectangle at the zero-end of the ruler and counting to seven (not eight) and ten (not eleven).

Part 4. Finding the Area of a Right Triangle Made from Half a Rectangle

Class time was running out, so we did only one more quick activity. The instructor held up a three-inch by five-inch rectangle. "Here is my rectangle named 15. Its area is fifteen square inches. I am going to draw a line on the rectangle from one corner to the opposite corner. And now I will cut along the line. What do I get?" She held up the two resulting triangles, and showed how one fit on top of another. "Two triangles. And they are alike. OK then, here is the big question. How many square inches are in each triangle? What name should I give each triangle?"

Some children shouted, "You take half."

"Half of what?"

"Half of fifteen."

"How do I do it using the calculator? Do you remember last week's lesson with clay?"

"You divide."

"What do you divide?"

Children were not sure of the answer here. Some said, "You divide by three." Others said, "You divide by five." Eventually several got it: divide by two. The teacher wrote on the board 15 ÷ 2 =.

"Try it on your calculator." Children tried it and got 7.5. Several read this as seventy-five. "It is seven point five. That is halfway between seven and eight square inches. We will learn more about this in another lesson."

Several children asked to take their rectangles home.

REMARKS

Possible Extensions

(1) After children have named their rectangles by their areas, have them line them up according to size. Here they can see that a one-inch by ten-inch rectangle has the same name (area) as a two-inch by five-inch.

(2) Have children draw diagonals on some of their rectangles and cut them into two right triangles and find the area of each. Because each triangle is one-half of a rectangle, we can compute its area by dividing the area of the rectangle by two.

(3) What if you made a rectangle twenty-three inches by fifty-seven inches and cut it into two triangles? (Show about how big that would be.) What would the area of each triangle be?

(4) (Hard problem) If I want to make two triangles, each with an area of twelve square inches, how can I do it? I can start by making a rectangle. How big does the rectangle need to be? (Answer: 24 square inches.) Can you find more than one way to do it?

Notes

As mentioned above, giving each child a square inch cut out in paper of contrasting color was a key. It allowed them to count accurately.

It is important to say that we are finding the areas (first of the rectangles and then of the triangles), and that we are naming the figures using the areas as the names. Using mathematically correct terminology is important.

CHAPTER 16

Ants' Roads

This lesson has been taught in a number of second-grade classrooms. In the lesson children measured and computed distances in centimeters and millimeters. The tasks were presented in the form of stories about roads traveled by ants.

PROPS

- one meter stick
- calculator poster
- picture of Ants' Roads for each child (see end of chapter). It has five points connected by five straight segments. Each point is marked with a small icon: house, butterfly, shell, flower, and grasshopper. We designed the picture so that one of the measurements (home to butterfly) was a precise number of centimeters. The others involved millimeters as well.
- ruler with centimeters and millimeters for each child
- blank sheet of paper, pencil, and calculator for each child

THE LESSON

"Do you know what an ant is? Do you know where it lives? Have you ever noticed how ants make narrow roads when they go from one place to another? Here we have a make-believe picture of some roads that some ants built. We are going to see how far they travel when they go from one place to another. First let's see if we can name all the places that are on the picture. Let's name the pictures. Where shall we start?" Our groups started with the house, which they called house or home. They were able to name all the icons on the picture except the shell; some called it a snail.

The instructor drew on the board a facsimile of the picture, including icons. She referred to it throughout the lesson. (An overhead transparency could be used here.)

"What do we know about rulers? Let's look at the side of the ruler that has centimeters. That's the side with the numbers one through thirty. How long is a centimeter?" Here the instructor drew some straight lines on the board and asked whether they were one centimeter in length.

The instructor named the traveling ants after children in the classroom. "Suppose Billie Jo wants to travel from home to the butterfly. How far does she walk? Let's measure and find out."

For some children, this was the first time they had used a ruler to measure. We found that

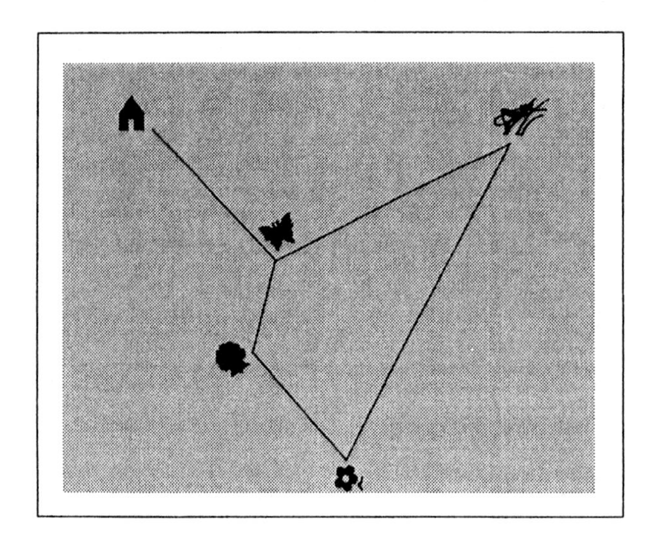

ILLUSTRATION 12. A picture of ants' roads.

they would line up their ruler so that "1" was at the home, rather than the end of the ruler. We simply told them to put the end of the ruler at home.

Children began to call out the number nine.

"Nine what?"

"Nine centimeters."

"What does it mean?"

"Billie Jo will walk nine centimeters from home to the butterfly." We wanted at least one child to verbalize this whole statement. Children were told to write on their blank sheets of paper:

home, butterfly 9 cm

This was very difficult for early second graders. They could print "home" fairly easily, but printing "butterfly" was a big task. When we do this lesson again, we might choose icons whose names are short and easy to spell, such as bug, leaf, rock, and nut.

The instructor told the children that "cm" was an abbreviation of the word centimeters and that scientists use it rather than writing out the whole word. "Now Billie Jo continues her walk. How far does she travel from the butterfly to the shell?"

The correct answer is between four and five centimeters. Children can get this. We asked them to notice the little lines between four and five on the ruler, and the spaces between them. "How many lines and how many spaces are there?"

"Nine lines and ten spaces."

"What does each little space measure?"

"One millimeter."

"So we can say Billie Jo walks 4 cm and how many millimeters when she goes from the butterfly to the shell?" We got disagreement here; some kids said 6 mm and others said 7. The instructor said, "Well, I have to record something on the board, so let's vote about which one you prefer." The children chose 4 cm 6 mm.

Write on your sheet:

butterfly, shell 4 cm 6 mm (or if you prefer, 4 cm 7 mm)

Just as you put a decimal point between the number of dollars and the number of cents when you are dealing with money, so you can put a decimal point between the number of centimeters and the number of millimeters in measurement.

Scientists write:

butterfly, shell 4.6 cm

So you can write on your sheet:

butterfly, shell 4 cm 6 mm or 4.6 cm

We found children had no difficulty accepting this. But they needed to read it aloud as four point six centimeters, or four centimeters and six millimeters, several times.

"Billie Jo continues her walk. How far does she travel from the shell to the flower?"

Again, there was disagreement: seven centimeters two millimeters, or seven centimeters three millimeters. We voted, and seven centimeters two millimeters won. So we recorded:

shell, flower 7 cm 2 mm or 7.2 cm

"Billie Jo has two more places to visit. How far does she walk from the flower to the grasshopper?"

flower, grasshopper 17 cm 7 mm or 17.7 cm

Again, we voted to decide between 17.6 and 17.7 cm.

"Finally, how far does Billie Jo walk from the grasshopper to the butterfly?"

grasshopper, butterfly 13 cm 3 mm or 13.3 cm

The instructor had been neatly writing on the board:

home, butterfly	9 cm
butterfly, shell	4 cm 6 mm or 4.6 cm
shell, flower	7 cm 2 mm or 7.2 cm
flower, grasshopper	17 cm 7 mm or 17.7 cm
grasshopper, butterfly	13 cm 3 mm or 13.3 cm

"Now Matt decides to take a walk. He starts at home and goes to the butterfly and then to the shell. How far does he travel?"

The kids' answers to this question were unexpected. They had no idea that they could find the answer by adding 9 and 4.6. Several used their rulers again, measuring from home to the butterfly, and then attempting to pivot their rulers by holding a point at the butterfly. Of course this would work if they could hold it precisely, but they could not. The instructor asked them to look at the numbers on the board and think about how they could use their calculators to get the answer. "We know that the distance from home to the butterfly is 9 cm, and the distance from the butterfly is 4.6 cm. What do we do with those two numbers to find out how far Matt walks from home to the shell?" They had learned to associate a number with a line's length, but they had not learned that you could associate the sum of two numbers with the total length of two segments. Finally one child said, "You plus them!" The instructor wrote on the board [9][+][4.6][=] and using the calculator poster, she pointed to the keys in sequence.

"What is the answer? What do you see on your display?" Some read the answer as 136. The instructor asked them to look closely and notice the decimal point between the 3 and the 6, and try again. Some could read the display correctly.

"So the answer is 13.6. Of what? 13.6 centimeters. So what does it mean?"

"When Matt walked from home to the shell, he walked 13.6 cm." Several could verbalize this. We consider it very important.

"Matt continues his walk to the flower. How far does he travel in going from home to the butterfly to the shell to the flower? Use your calculator." There were several solutions here:

[9][+][4.6][=][+][7.2][=] Display: 20.8

[13.6][+][7.2][=] Display: 20.8

"What does it mean?" Again, we wanted at least one child to say, "It means Matt walked 20.8 centimeters when he walked from home to the flower."

"Now Ryan decides he wants to visit Matt at the flower. Ryan starts from home and walks to the butterfly. Then he decides to take a shortcut and go straight from the butterfly to the flower. How far does Ryan walk? First, you need to take your ruler and draw a straight line from the butterfly to the flower."

The instructor drew such a line on the picture on the board. Children had difficulty here. Very few could actually draw a straight line using their rulers. "How long is the straight line from the butterfly to the flower? Measure it!" We voted, and 10.4 cm was decided on. "So how far does Ryan walk from home to the flower, taking his shortcut?"

[9][+][10.4][=] Display 19.4

"What does it mean? Ryan walks 19.4 cm from home to the shell, and Matt walks 20.8 cm from home to the shell. Who walks farther, Matt or Ryan?"

"Matt!" The children had no difficulty with this.

"How much farther? How can we figure out how much farther Matt walked than Ryan, using our calculators?"

This was extremely difficult for children. They had no idea that 20.8 − 19.4 would provide the answer. The teacher tried to help by giving a similar situation with positive integers which she knew they were familiar with: "Monica has eight apples, and Marcus has five. How many more does Monica have? What do you do to find the answer?"

One child said, "You take away."

"What do you take away?"

The instructor wrote:

[20.8][−][19.4][=] Display 1.4

"What does 1.4 mean? 1.4 of what?"

A girl finally said, "It means that Matt walked 1.4 centimeters farther than Ryan." The instructor showed, using a meter stick, how far 1.4 cm was (not very far).

Last question: "Julie decides to take a stroll all the way around the map. She starts from home and walks to the butterfly and then to the grasshopper and then to the flower and then to the shell (no shortcut) and then to the butterfly and then back home. How far does she walk? What do we have to add?"

[9][+][13.3][+][17.7][+][7.2][+][4.6][+][9][=] Display: 60.8

We waited until every child was able to get 60.8 on the display. "What does 60.8 mean?"

"It means that Julie travels 60.8 cm if she walks all the way around the map, beginning at home and ending at home."

Using the meter stick, the instructor then showed the children how far Julie would have walked if she had walked in a straight line.

One boy said, "Wow! 60.8 cm! That's a mile for an ant!"

REMARKS

As mentioned above, we might choose icons with shorter names. The lesson could be broken into two sessions.

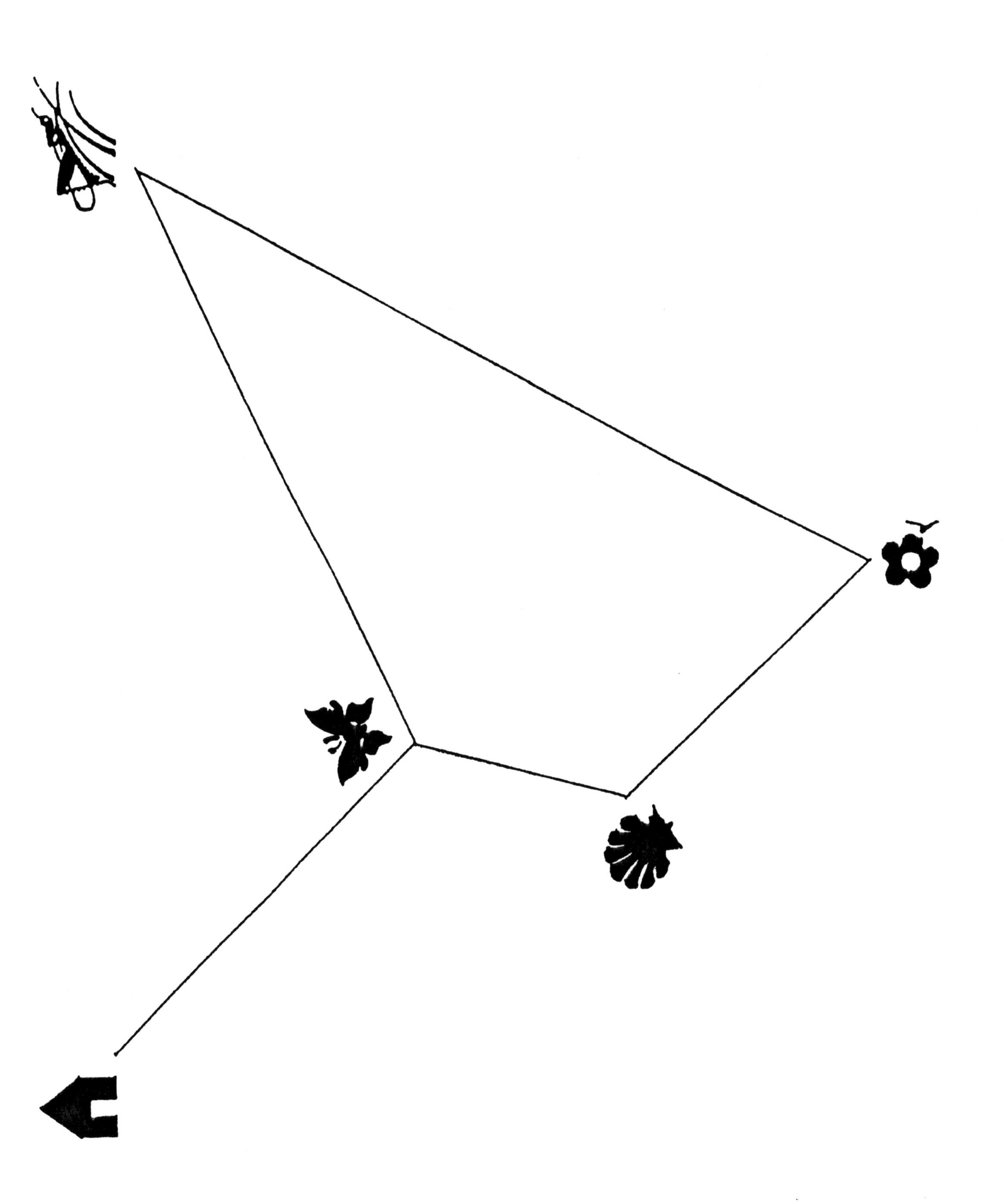

CHAPTER 17

Spider Web

This lesson has been taught in a number of second-, third-, and fourth-grade classes. The description below is from a second-grade class. It is typically taught right before Halloween. Several teachers have embellished it by reading *Charlotte's Web* and including a science unit on spiders. At the end, it is also an art lesson (see Illustration 13).

PROPS AND TOOLS

- a sheet of paper with a picture to work with (see end of chapter) for each child
- one piece of paper to write on (We used half sheets; some children will write right on their picture.)
- a pencil, ruler, and calculator for each child
- a large calculator poster visible to all in the classroom
- two meter sticks (for measuring how much silk the spider spins)
- a piece of chalk tied to a string, for drawing a large circle on the board
- a drawing of a finished web on one of the pictures, with a black plastic spider glued to the center of the web (see end of chapter)
- a package of plastic spiders to pass out to each child at the end of the lesson (optional)

Description of the Picture

On the sheet of paper, there is a circle, approximately 15 cm in diameter. On the circle there are six clearly marked points forming vertices of a regular hexagon. (The hexagon itself is imaginary; its sides are not drawn.) The hexagon is not aligned with the sides of the sheet of paper; it is in an ''arbitrary'' position. The interior of the circle is empty, but the exterior is embellished, showing some grasses growing around the pipe.

THE LESSON

''Today we are going to pretend we are spiders! Do you know what a spider is? Can you describe it?'' We learned that a spider has eight legs (It is not an insect!) and the one outstanding feature of most spiders is that they spin webs.

''Why do they spin them?''

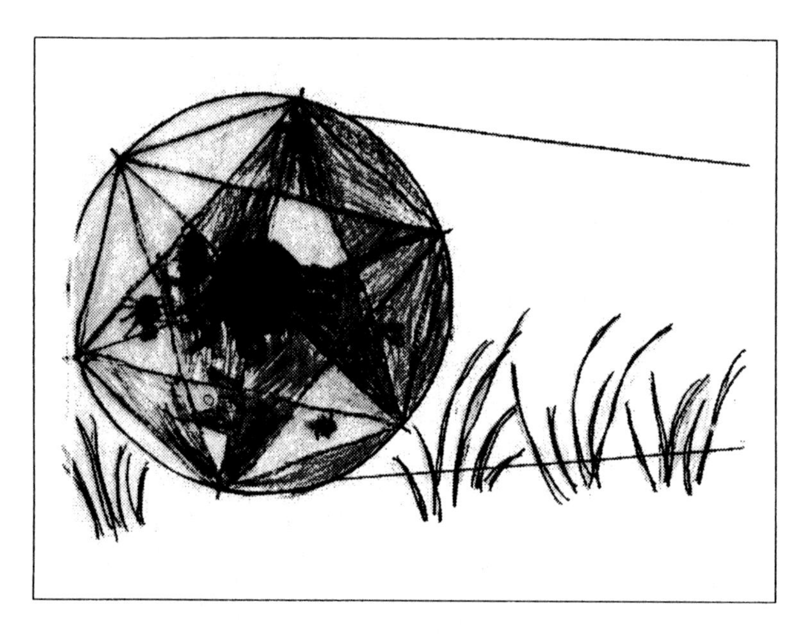

ILLUSTRATION 13. A third grader's spider web drawing.

"To build a trap to catch food."

"What do they do after they catch an insect or other bug?"

"They spin thread around it to hold it, and then they suck out its blood!"

"How do they spin the web?"

"They have spinnerettes in the bottom of their abdomens, and the spinnerettes produce the thread."

"Is the thread strong enough to sew with?"

"No!"

Part 1

"We are going to be spiders, and we are going to spin a web. Do you have a picture? It shows a piece of old water pipe lying in a meadow. Don't draw anything yet, but here is what we are going to draw." I held up the completed drawing of the web, with the black plastic spider in the middle. Then, using a piece of chalk tied to a string, I drew a large circle on the board. Using as a distance between points the radius of the circle, I marked off along the circumference of the circle six clearly marked points.

"The first thing a spider has to do when spinning a web is to spin some thread clear across where she wants to make the web. Are you ready to spin? Watch me! I will show you how we are going to make our webs, and since you can't spin a web, you will draw it. Later we will find out how much thread we spiders used for our webs. Ready?

"First, we want to put a long thread straight across our pipe. I will use a meter stick. I put it down so that I can draw a straight line between this point and this one." I picked two points on either side of a diameter. "See how I have to hold the ruler still while I draw the line? And see how I put my chalk right up against the ruler? Try it. Hold your ruler steady and put your pencil lead right up against the ruler. Have you got one line drawn across the pipe?

"Now I am going to spin another thread across the pipe, just like the first one." I picked two other points on either side of a diameter. "Can you spin another line across your pipe?

"Now let's spin a third thread clear across our circle." I picked the final two points on either side of a diameter. "Wow! Look at this! My three lines almost meet in the center. Do yours almost meet?

"Are we tired of spinning yet? We are not nearly done! Let's write down on our half sheets of paper: 3 threads." I wrote this in large letters on the board.

"Now let's measure how long one of our threads is. Use the centimeter side of the ruler—the one that goes up to thirty-one. How many centimeters long is one thread?" Children had some difficulty here. We designed our drawing so that the circle's diameter was almost exactly fifteen centimeters. But several children put the point marked "1" on the ruler on one point and claimed that the length was 16 cm. The teacher and I walked around to be sure that each child began measuring from the end of the ruler.

"Do you think that the other two threads will be the same length?" Kids knew that they would: They go from one side of the circle through the center to the other side. "Now let's add to our paper":

3 threads 15 cm each

I also wrote this on the blackboard.

"Are you ready to spin some more? Watch me! It's going to get more complicated now. Now I am going to connect every second point. So I start here, skip one, and end here. Do you see how I am doing it? You can try it. Now I start where I just ended, skip one, and end here. Do you see? Try it. Finally, I start where I just ended, skip one, and end where I first started." I drew three short diagonals. "Do you see what shape I made? A triangle! So I made three more threads. Do you have three more threads? Let's write on our paper again: three threads. How long are they? Can you measure one?" Children for the most part got 13 cm here. "Are all three the same length? We can check it. Yes, they are. So let's add to our writing":

3 threads 13 cm each

I wrote this on the board.

"We are getting to be tired spiders. But we have a few more things to spin. Are you ready? Watch me. Again I am going to connect every second point, making another triangle." I used a different color chalk here, to make the second triangle more noticeable. "Can you do it? Try it!

"Do we know how long the threads are? Are they the same as before? Let's check. Write again":

3 threads 13 cm each

I wrote this on the blackboard.

"We only have one more thing to spin. Watch me. We are going to put six short threads all around. Can you do it? Try it.

"Now let's measure them." The actual measurement was very close to seven centimeters and five millimeters. Children had not learned about millimeters before. I explained.

"When we measure our six short lines, we see that they come out between seven and eight centimeters. Do you see the small black lines between the seven and eight on your ruler? Each small space is a millimeter. Can you count how many millimeters there are in a centimeter? Ten. When we measure our lines, we see that we are between seven and eight centimeters. Can you count how many millimeters we go beyond seven? We go five. So we want to write seven centimeters and five millimeters. Scientists do this in a fast way. They write 7.5 cm. They put a decimal point between the number of centimeters and the number of millimeters. So write on your sheets":

6 threads	7.5 cm each

"I know we are all tired. Let's draw a spider sitting on the web."

Part 2

"Now let's see if we can figure out how much silk we spun."
Each child was supposed to have on his/her piece of paper:

3 threads	15 cm each
3 threads	13 cm each
3 threads	13 cm each
6 threads	7.5 cm each

"We first drew three lines across the pipe. How long was each one? Each was 15 cm. How many centimeters is that altogether? Can we use our calculators?"

[3][*][15][=]

or

[15][+][15][+][15][=]

45 cm. Let's write:

3 threads	15 cm each	45 cm
3 threads	13 cm each	
3 threads	13 cm each	
6 threads	7.5 cm each	

"Now what about the triangles? How much did we spin there?"

[3][*][13][=] 39 cm

"That goes for both triangles, right? So let's write":

3 threads	15 cm each	45 cm
3 threads	13 cm each	39 cm
3 threads	13 cm each	39 cm
6 threads	7.5 cm each	

"Finally, we need to compute how long our last six threads were":

[6][*][7.5][=] 45 cm

"Let's write":

3 threads	15 cm each	45 cm
3 threads	13 cm each	39 cm
3 threads	13 cm each	39 cm
6 threads	7.5 cm each	45 cm

"Now let's add up all the numbers":

[45][+][39][+][39][+][45][=] Display: 168

"Each of us spiders used 168 cm of thread. (It's a lot!) Let me show you with two meter sticks how much that is." I held up two meter sticks end to end, and showed where 168 cm was.

Although time did not permit in this class, children can now color their webs.

REMARKS

This lesson is rather complicated, and we recommend that you draw and then measure and record after each different part of the web is drawn, as we have done above. We recommend that children write on their sheets as we have demonstrated. Otherwise the measuring can get out of hand, and the children can get confused.

It is optional in this lesson to use the words "diameter" and "hexagon." We elected not to.

At the end of the lesson we invited the children up to measure themselves with the meter sticks and compare their heights with the amount of silk that they had "spun."

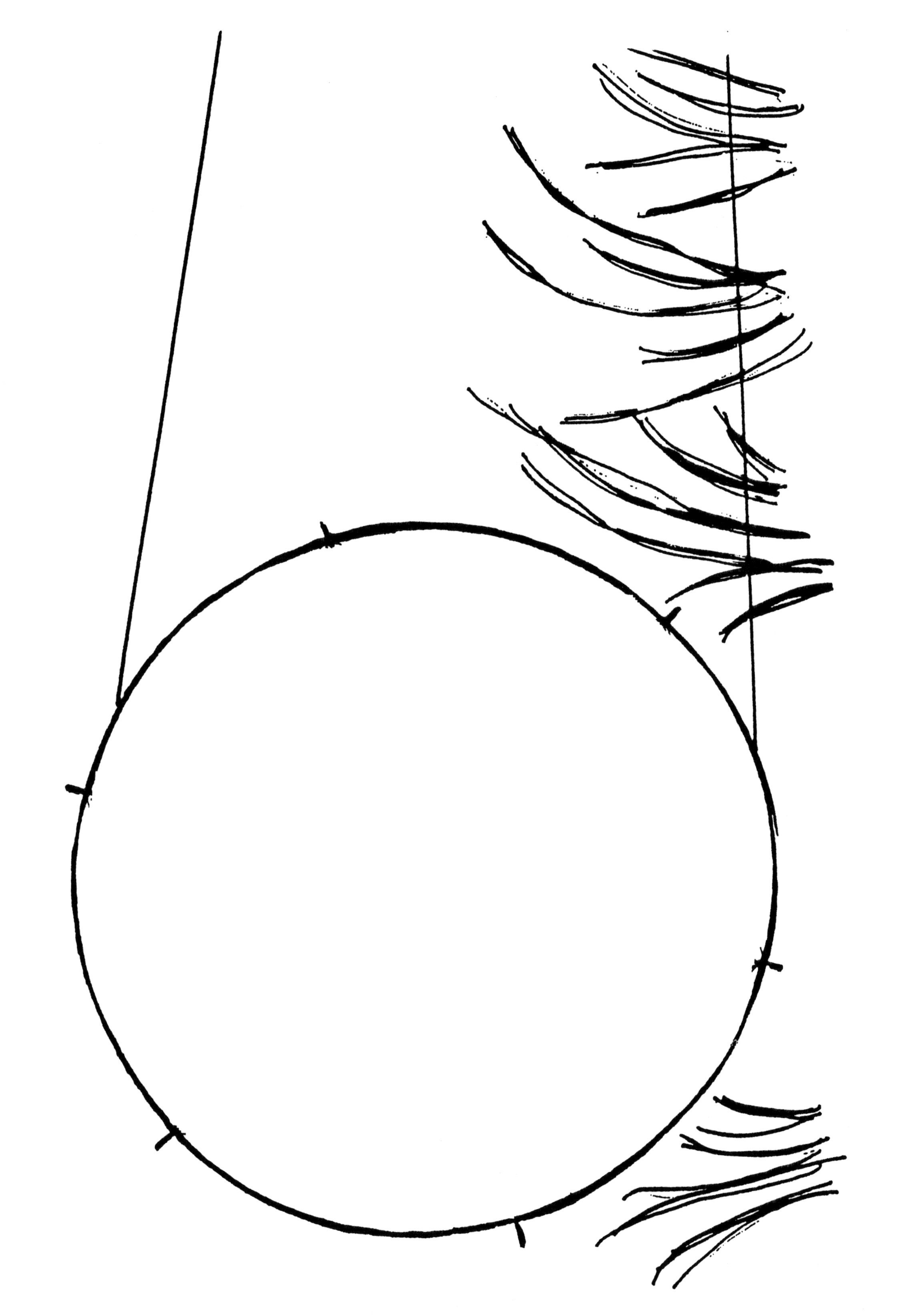

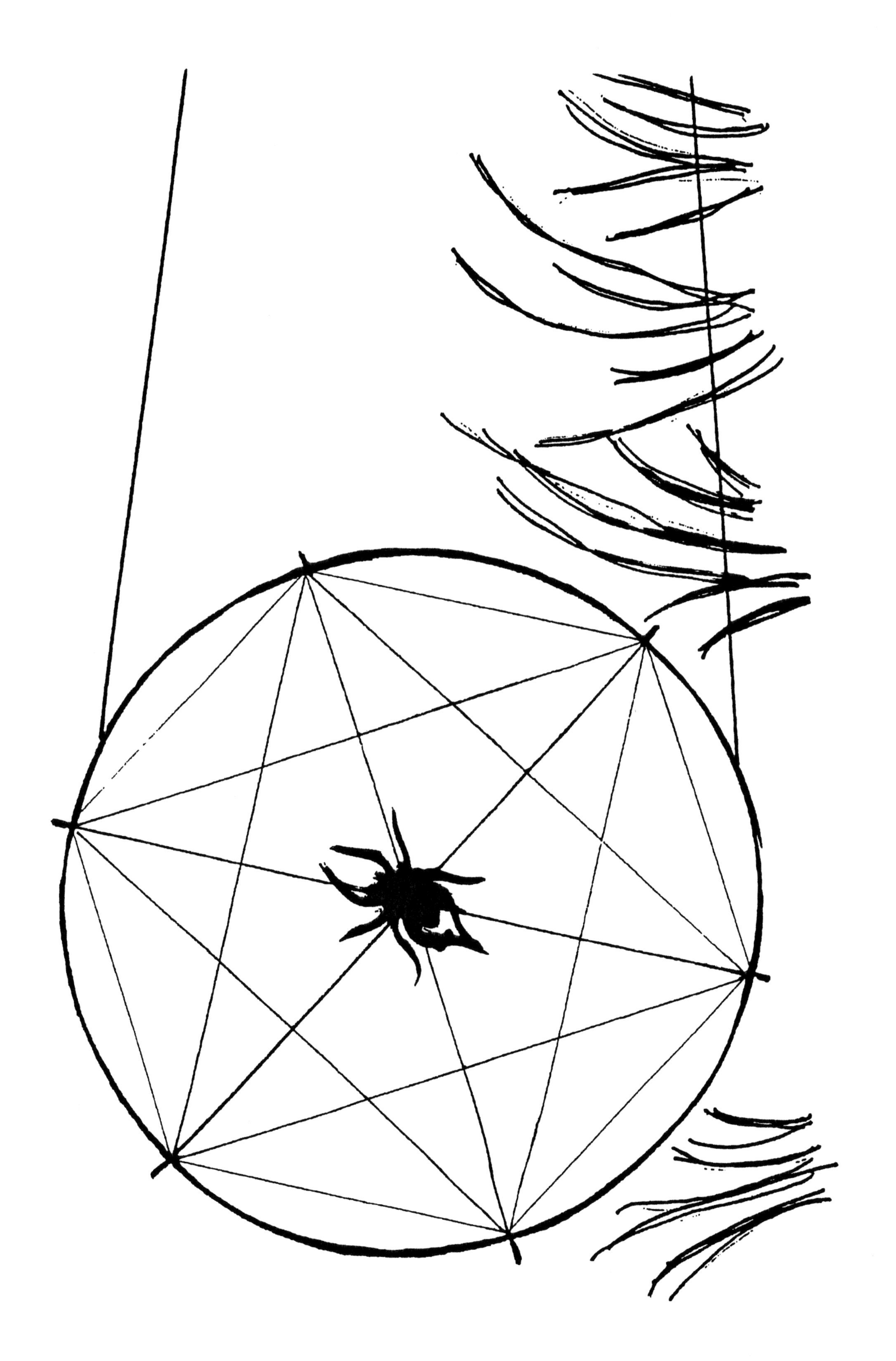

CHAPTER 18

Traveling Bugs (Grades Three and Up)

One interesting (and in general difficult) problem is the "Traveling Salesman" problem. It says: given a set of points, find the shortest path that goes through all the points and ends in its starting point (the path is circular). The name comes from the following interpretation. You can think of the points as cities to be visited by a traveling salesman who wants to find the shortest route for himself. In general, additional constraints are added, and they indicate between which points you can or cannot make a direct connection.

PROPS AND TOOLS

- a sheet of paper (with no lines or grid) with a set of six to twelve points clearly marked in an irregular pattern (see end of chapter)
- a ruler, pencil, eraser, and calculator, one per child
- an overhead transparency of the picture
- a transparent ruler (for use on the overhead)

Children may work in groups, but each child should have his/her own set of tools.

THE LESSON

Each point is a home of a little bug. Bugs like to visit each other and want to build a path that goes around and joins all their houses. (They will also jog on it.) They want to make the path as short as possible. Different bugs propose different paths. Which one is the shortest? (The teacher should display one path that is clearly not the shortest.)

Activities

DISCUSSION OF THE STORY

- This is a "make-believe" story. Why?
- Would we like to add something to make it more interesting? (Yes.)
- How are we going to find out the shortest path? Everybody will draw his/her path and measure its length, and we'll compare the results.
- Which solutions are acceptable? The path must be circular. Does it have to consist of straight segments? No, but the shortest path consists of straight segments. (This may be a difficult point.) Can the path cross itself? Yes, but the shortest one doesn't. (Another difficult point.)

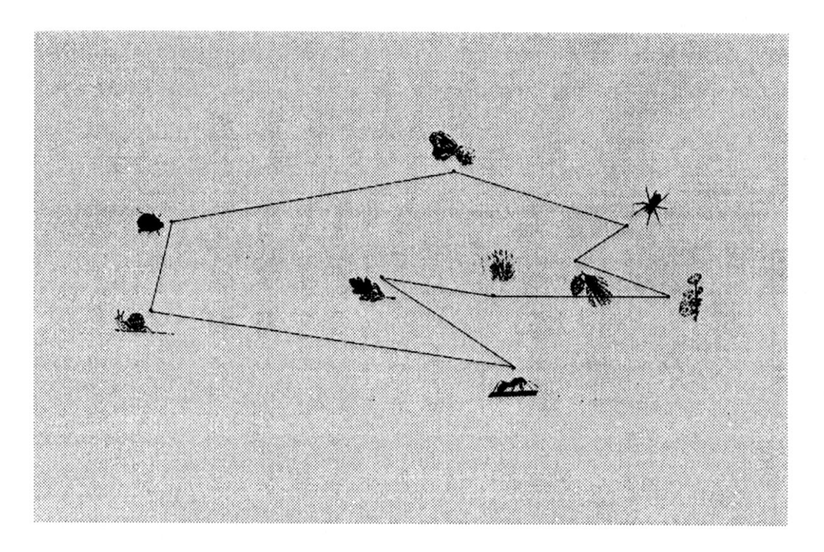

ILLUSTRATION 14. One possible path for traveling bugs.

DOING THE TASK

Children draw their paths. They measure the lengths of the segments in centimeters and millimeters and compute the total. On the board the teacher can keep a list of children's names and the lengths of their paths.

DISCUSSION OF RESULTS

Check if the proposed paths are circular, that is, if each one goes through all the points and ends at the starting point. The best solution(s) are chosen. Measurements and calculations are checked. The teacher can draw on the transparency the path that is reported to be the shortest, and children can measure and vote on the length of each segment. Is it really the shortest? If another path appears to be shorter, the teacher can draw it, using a different color marker, and drawing only the segments that are different from the first path. The question is posed: Do we know that the path(s) we found is (are) the shortest? Likely answer: no, maybe there is a better one. Children should be encouraged to give arguments that their solutions cannot be improved.

CHILDREN DRAW BUGS AND COLOR THEIR MAPS

Example:

. d

. h

. a . f

. b

. c

. i . e

. g

Proposed paths:

- adhegbfcia
- adfbhegica
- icfgbehdai

Notice that the way the points are labeled does not suggest how to choose a path.

CHAPTER 19

Boxes: Surface Area and Volume

This lesson has been taught in grades two to five. The description below is from a second-grade class. The lesson is lengthy and could be broken into two sessions.

The task is to measure and calculate the surface area and the volume of a cardboard box. Older children will also cover the box with square-inch graph paper. Our students had had several lessons on measuring lengths, and one or two on computing the area of a rectangle (such as "Rectangles").

PROPS AND TOOLS FOR EACH CHILD

- a folded cardboard box. We used flattened gift boxes from a department store, purchased for 25¢ each; they were dark green and 6 in. by 6 in. by 4 in. (other sizes will work as well).
- a ruler
- a piece of blank paper
- a pencil
- a calculator
- six post-it notes (we used those that were 3 in. by 3 in.)
- a one-inch square piece of paper
- for older grades, three 8-1/2 in. by 11 in. sheets of square-inch graph paper (we used pink paper)
- a glue stick or paste

Other props include a calculator poster put in view of all students; an unfolded box, "papered" on all six sides (and on two sides on the inside of the box) with one-inch graph paper; six post-it notes, four labeled 24 sq in. and two labeled 36 sq in.; a 21 × 8 sq in. piece of graph paper (with grid plainly visible; this is exactly enough paper to "wallpaper" the box); some sheets of blank paper; wide clear tape to repair torn boxes (narrow scotch tape or glue will also work); at least eight cube blocks (one-inch cubes).

THE LESSON

"Does everybody have a ruler, pencil, blank sheet of paper, and calculator? How about a square inch? And six little note sheets with sticky backs? And three sheets of pink graph paper? And a flat box?

"Today, we are going to make a box and measure its surface area and volume. But first

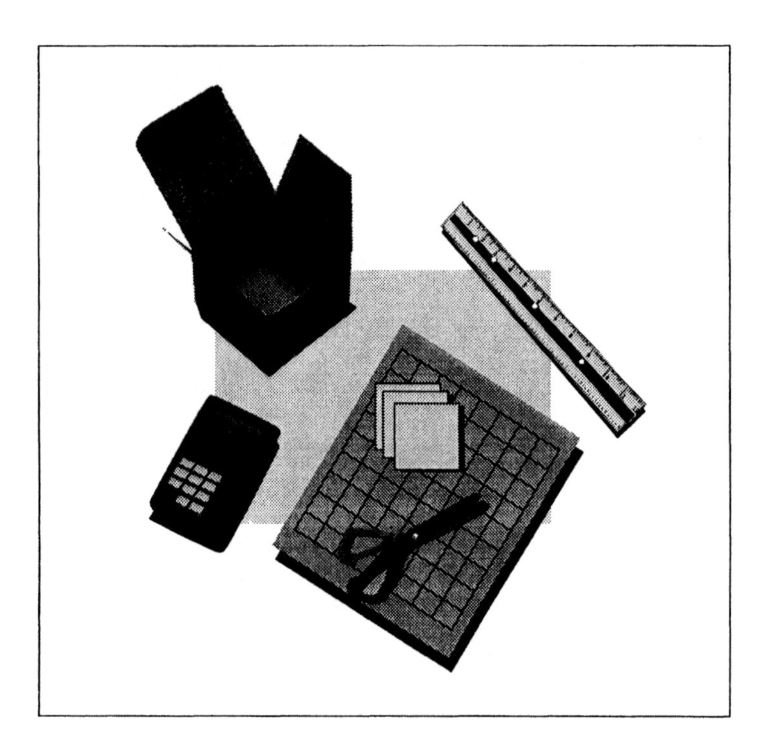

ILLUSTRATION 15. Some supplies needed for boxes lesson.

let me show you how to make the box. Hold it so that the grey part is toward you, like this. Now press on both edges, and it will unfold like this. Press the bottom down, and fold the flap for the top, so you can close the lid."

Many children did not wait for these instructions. The box is rather tricky to make, and some children tore the bottoms of their boxes. We made repairs using a piece of wide tape stuck on the bottom of the box.

"First, let's look at our box. How many corners does it have? Can you point to the corners?" The children first said there were four corners, then six. Finally there was agreement that there are eight. "Will you point to each corner in turn and count with me? One, two, . . . eight.

"What about edges of the box? What is an edge? How many are there?" These were more difficult to count. Children first thought there were eight. But again, finally there was agreement that there are twelve. "Let's count the edges. There are four around the top, four around the bottom, and how many that go up and down? Four. What is 4 + 4 + 4? 12! Another way to do it is three fours, or 3 times 4. You can try that on your calculator: [3][*][4][=]. What do you get? 12. The box has twelve edges."

The instructor then pointed to a face of her box. "What is this called?"

Children said, "A side."

"Yes, you can call it a side. But it also has another name. What is this part of my body?" She indicated her face.

The children responded, "Your face."

"The sides of a box are also called faces of the box. How many faces does a box have? Can you count them?" Children counted these accurately: six faces.

"Now we know a lot about the box. It has how many corners?"

"Eight."

"How many edges?"

"Twelve."

"How many faces?"

"Six."

"Already you are getting to know your box!

"Here is our first big task for today. Do you know what wallpaper is?" Children know this! "We want to decide exactly how much wallpaper we need to cover our box completely. That means we want no dark green to show. How are we going to do it?" We discussed this. Children were not very sure about what to do.

"We need to find how much paper we need for each face, and then we can add these amounts to get the total we need.

"But how can we know how much paper we need for a face? We have to do some measuring. If we can find out how tall and how wide a face is, we can figure out its surface area, or how much paper we need for that face. So let's start by measuring the edges. Let's start at the top of the box. How long are the edges around the top? Take your ruler and find out. Use the side of your ruler with inches, the one that goes up to 12. We will record what we find on our blank sheets." She tacked on the board a sheet of paper.

Some children needed help here, but there was general agreement that each edge around the top was six inches long.

"Here is what we want to write on our data sheets":

edges	
how many?	how long?
4	6 in.

"Scientists like to use abbreviations; they don't write out the whole word, 'inch'; instead, they write 'in.' You can be a scientist today and just write 'in.' " The instructor wrote this on the data sheet, using a broad marker. She also wrote it on the board, in large print. She and the teacher went around to each child to be sure he or she had begun a data sheet correctly.

"Now let's measure the edges around the bottom. What do you get? How many are there, and how long are they? Here is a challenge. Can you tell without measuring?" Some children responded that there would be four again, and each would be six inches long. "Why?"

A few children said, "The top and the bottom of a box are the same size!"

We wrote further on our data sheets:

edges	
how many?	how long?
4	6 in.
4	6 in.

"Now what about the other edges, the ones that go up and down? How long are they?" Children were able to measure and say they were four inches long.

We added to our sheets:

edges	
how many?	how long?
4	6 in.
4	6 in.
4	4 in.

"Now we are ready to compute the area of each face. What do you think the area of the top face is? What do we know about it? It is six inches wide and six inches high. We want to know how many square inches it is." She displayed her "papered" box, showing the top.

"We have how many rows of how many square inches? You can use your square inch to get an idea. Can you see that there are six rows of six?" She showed this, using her square inch. "We can write this 6 × 6. Let's write on our data sheet":

edges	
how many?	how long?
4	6 in.
4	6 in.
4	4 in.

faces	area
top	6 × 6

"What are six sixes? We can do it with the calculator: [6][*][6][=]. How many square inches? Let's write":

edges	
how many?	how long?
4	6 in.
4	6 in.
4	4 in.

faces	area
top	6 × 6 = 36 sq in.

She explained that scientists abbreviate square inches as "sq in."

"Let's take a little yellow sticky note and write 36 sq in. on it, and stick it on the top of our box." This was a very good thing to do; the children even said, "Oh, that will be a good reminder."

"Now let's calculate the area of the bottom of the box. How many square inches will it be?" Some children knew that we don't even need to calculate; we already know. The area is 36 sq in. We discussed why we know this without calculating. We wrote:

edges	
how many?	how long?
4	6 in.
4	6 in.
4	4 in.

faces	area
top	6 × 6 = 36 sq in.
bottom	6 × 6 = 36 sq in.

"OK, let's label that face too with another sticky note: 36 sq in. Now we are done with two faces. We have how many more? Four! So we need to find their areas. Let's start with this one. How wide is it, and how tall?" Some children had to remeasure; the instructor said they could also tell by reading from their data sheets. "The face is 6 in. wide and 4 in. tall. Let's write":

edges	
how many?	how long?
4	6 in.
4	6 in.
4	4 in.

faces	area
top	6 × 6 = 36 sq in.
bottom	6 × 6 = 36 sq in.
side	4 × 6 = 24 sq in.

"Let's label the side with a sticky note: 24 sq in.

"What about the other sides?" Again the instructor showed her "papered" box. Children responded that the other sides were the same size: 24 sq in. We wrote on our data sheets:

edges	
how many?	how long?
4	6 in.
4	6 in.
4	4 in.

faces	area
top	6 × 6 = 36 sq in.
bottom	6 × 6 = 36 sq in.
side	4 × 6 = 24 sq in.
side	4 × 6 = 24 sq in.
side	4 × 6 = 24 sq in.
side	4 × 6 = 24 sq in.

"Let's label the rest of the sides with sticky notes." She did so with her box. The children seemed to particularly enjoy being able to keep their places with the sticky notes.

"OK, we are about ready to say how much wallpaper we need to paper the box exactly. How can we figure it out? What do we know, and what do we need to do?" Only one or two children were able to verbalize what we needed to do: add up the areas from all the faces.

"Here is one way to do it with the calculator":

[36][+][36][+][24][+][24][+][24][+][24][=]

"Try it." She showed the keystrokes using the poster. "What do you get? 168. 168 what? 168 sq in." She and the teacher walked around to help each child get 168 on his or her calculator display.

"Let's write on our data sheet":

edges	
how many?	how long?
4	6 in.
4	6 in.
4	4 in.

faces	area
top	6 × 6 = 36 sq in.
bottom	6 × 6 = 36 sq in.
side	4 × 6 = 24 sq in.
side	4 × 6 = 24 sq in.
side	4 × 6 = 24 sq in.
side	4 × 6 = 24 sq in.
total	168 sq in.

"Would you guess that the area would be so big?

"Here is a challenge question. I have a sheet of wallpaper here. As you can see, it is twenty-one inches long and eight inches wide. I am wondering if it is big enough to paper the box. Can you find out? What do you have to do?"

[21][*][8][=]

"168! There are 168 square inches of paper here. So is it enough? Yes, exactly." She then folded the paper around the box, to give an indication that there was enough paper.

"There is one more thing we want to do. We know that the box has surface area. But it has something else too: volume. Volume is the amount of space occupied by the whole box. It consists of empty space inside and part of the space occupied by the cardboard itself. So the amount of stuff you can put inside is slightly smaller than the volume of the whole box. Can we figure out the volume of the box?

"Here are some inch cubes." She used only twelve. "Scientists call them cubic inches! I will start to stick them in the bottom of the box." She opened the box and showed the inside. Two of its "walls" were papered with one-inch graph paper. "I will have to stretch the box just a little bit to get them to fit. How many can I fit in the bottom? Six in a row. Then six more. Then. . . . How many rows of six?" Children knew there were six rows. "So how many inch cubes can I put in the bottom layer? [6][*][6][=]. Thirty-six cubic inches. And how many layers?" Children could see there were four in all. "So, if I stretch the box a bit, how many cubic inches will it hold? [36][*][4][=]. 144 cubic inches.

"Let me show you one last thing before we quit. Go back to your data sheet. Do you see how we recorded the number of edges and how long they were? There is another way we can compute the volume of the box. We can multiply together how long the box is (6 in.) by how deep it is (6 in.) by how high it is (4 in.). Try it: [6][*][6][*][4][=]. 144 cubic inches!"

We collected all props. However, children wanted to keep their boxes and their data sheets. If possible, we recommend arranging for this. In the second grade described here, the lesson was taught a few days before Mother's Day. The children kept their boxes, made pictures to decorate the sides with, and filled them with trinkets for their mothers. We got tissue paper from the department store for lining the inside of each box.

REMARKS

When this lesson is taught in older grades (third, fourth, and fifth), children finish by using the pink graph paper to cover their boxes. They must figure out how big each side needs to be, cut out the right size for each side, and glue the paper onto the box.

CHAPTER 20

Popcorn (Grades Three or Four and Up)

The boxes lesson or something similar should precede this lesson.

Many nice lessons can be taught if, in a classroom, children have a large selection of boxes (any kind).

PROPS AND TOOLS

- breakfast cereal boxes (individual servings) and small sample toothpaste boxes for each group [Children may work in groups of four, so six (or more) boxes of the same size are needed. Small toothpaste boxes can be of different sizes and can even be made from construction paper by children.]
- cups, scissors, and plates
- a pound of uncooked popcorn
- construction paper, glue, and rulers
- optional: scales (They allow solutions not discussed here!)

THE LESSON

How many kernels of uncooked popcorn are needed to fill a cereal box?

Children can make guesses. Different methods may be proposed. The teacher may say that a specific method will be used this time.

(1) Each group gets a very small cup of popcorn.

(2) They count 100 (or 200) kernels.

(3) They construct, either from construction paper or by cutting a small toothpaste box to the right size, a container that holds exactly this amount of kernels.

(4) They measure the volume of this container, and of the cereal box, and estimate the amount of kernels needed to fill the cereal box.

After steps (1), (2), and (3) are finished, the children should discuss their methods of computation and a program that gives the answer.

Example

A small toothpaste box had a base 2 cm by 3 cm. When the height was cut to 2.5 cm, this small open container held 100 kernels of popcorn.

ILLUSTRATION 16. Props for popcorn lesson.

A cereal box with its top cut off was 2.3 cm by 8 cm by 12 cm. The formula estimating the number of kernels that the cereal box would hold: volume of the cereal box divided by volume of open toothpaste container times 100. The program:

[2.3][*][8][*][12][÷][15][*][100][=] Display: 1472.

($2.5 \times 2 \times 3 = 15$ was computed mentally)

The answer: approximately 1500 kernels.

CHAPTER 21

Sunflower Seeds (Grades Three to Four and Up)

This is a version of a very old problem.

Jim, Sue, Alice, and Toshiro bought one pound of sunflower seeds (One pound is a lot of seeds!) which they planned to share at a later time. They left the seeds at Jim's house. But Jim did not want to wait for his share, so he took one-fourth of the seeds. Soon afterwards, Sue came to his house and said to him, "I want my part now."

"Fine," he said. "Take your part." But he did not tell her that he had already taken his part, and she didn't notice it. So she took just one-fourth of what was left.

Just after she left, Alice and Toshiro came by Jim's house. "What's going on?" said Alice. "The jar is half empty!"

"Yeah," said Jim, "Sue and me already took our shares." (Children: what is wrong with Jim's grammar?)

So Alice and Toshiro divided the rest evenly between themselves.

QUESTIONS

Was the division fair for everybody? Did some children get more than others? Who got the most? Who got the least? Whose fault was it? How should it have been done? Can it be corrected?

Note: this problem can be solved with a calculator or mentally without any manipulatives. It can also be turned into a small play, with a pound of red beans substituted for seeds. In this case the answers to the questions should be a part of the play, with the audience playing the role of advisors. (Measuring cups or scales will be needed.)

ANSWERS

Let's compute how many ounces of sunflowers each child got. There are 16 oz in a pound. So Jim got $16 \div 4 = 4$ oz of the seeds.

He left $16 - 4 = 12$ oz of seeds. Sue took one-quarter of them. So she took $12 \div 4 = 3$ oz. What was left? $12 - 3 = 9$ oz of sunflower seeds. Finally, Alice and Toshiro shared this 9 oz, so each of them got $9 \div 2 = 4.5 = 4\text{-}1/2$ oz of seeds. So it was not fair. Alice and Toshiro got the most, and Sue got the least. It was Jim's fault. If he had told Sue when she came to his house that he had already taken his part, she would have taken one-third of the remaining 12 oz, which is 4 oz. The error can be corrected, if Alice and Toshiro have not eaten their seeds! Both Alice and Toshiro should give Sue one-half oz of the seeds, and then each will have 4 oz.

ILLUSTRATION 17. A balance scale can be used for the sunflower seeds lesson.

ILLUSTRATION 18. A spring scale can also be used.

CHAPTER 22

Cylinder

PROPS

- a cardboard cylinder from a toilet paper roll for each child
- a ruler for each child
- other tools and supplies, depending on the methods children decide on

THE LESSON

Children work in pairs. The task is to measure the surface area (side only) of the cardboard cylinder. There are several methods of measurement (children should be allowed to invent their own):

(1) Cut the cylinder, flatten it, and measure the area of the resulting rectangle. The shape is only an approximation of a rectangle. The height, measured in three places, was 11.2 cm, 11.3 cm, and 11.4 cm; the width was 13.7, 13.6, 13.8. The average height is 11.3 cm, and average width is 13.7 cm.

So the area is $11.3 \times 13.7 = 154.81$ sq cm (use a calculator). (Obviously the accuracy of the measurement was not very good, so the last digits are meaningless. To estimate accuracy, compute also the area based on smaller and larger values for the sides: $11.2 \times 13.6 = 152.32$ sq cm, and $11.4 \times 13.8 = 157.32$. So a more realistic answer is 155 sq cm plus or minus 2 sq cm.)

(2) Measure the circumference with a piece of string, then measure the string. Measure the height and compute the area.

(3) Put wet paint on the cylinder and make a print. Measure its area.

(4) Wrap the cylinder in a strip of paper cut to its size. Measure the area of the strip.

(5) Measure the cylinder's height and diameter. The height is 11.3 cm. The diameter measured in two directions gave 4.3 cm and 4.5 cm, giving an average of 4.4 cm. Multiply these numbers. How many times longer is the circumference than the diameter? The circumference is 13.7 cm. $13.7 \div 4.4 = 3.1136363$, a little more than 3, let's say 3.1. Multiply the height times diameter by a little more than three. $11.3 \times 4.4 \times 3.1 = 154.132$, rounded to 154 sq cm. This is pretty close to the previous value. So we have a new way of computing the area of a cyclinder.

ILLUSTRATION 19. One way to cut a cylinder in order to find its surface area.

REMARKS

Each pair should make their measurements, but one of the two cylinders should be left intact.

There are some key words which should be written on the blackboard. They should be listed in order of their occurrence in the lesson.

cylinder
surface area
rectangle
approximation
measure
height
width
accuracy
circumference
diameter

A word can be omitted from the list if children know its meaning and its spelling from some previous lessons.

CHAPTER 23

Fraction Bars

This lesson was taught in a grade four – five split classroom.

PROPS

- a calculator for each child
- six sheets of paper with centimeter grids (see Appendix for centimeter dot paper) [Each is approximately one-quarter of an 8-1/2-in by 11-in sheet. The sheets should be different colors. (We used red, green, yellow, orange, purple, and blue.)]
- two 8-1/2-in by 11-in sheets of white paper
- pencil, scissors, ruler, and glue (We used glue sticks.)
- an overhead calculator (if available), a transparency of centimeter dot paper, and a transparent ruler

The goal of this lesson is for each child to prepare a chart on which different fractions of one decimeter are represented by colored bars of appropriate lengths. Beforehand the teacher should prepare her own chart, to show to the children as a model (see Illustration 20).

THE LESSON

The children knew something was afoot because a table was prepared on which were laid out six stacks of different colors of centimeter grid paper and one stack of white paper. A box of glue sticks was also placed on the table. The children came up to the table, buffet style, and collected a sheet from each colored pile, two white sheets, and a glue stick. They were also asked to get a ruler, pencil, and scissors from their desks, and a calculator from the calculator caddy on the wall.

I had an overhead projector with calculator and my chart. I held up the chart and said, "We are going to make this today! Do you know what it is?" The kids described it as bars or strips of different colors, and they noticed that some bars were the same color. "You are right. The bars are of different colors and different lengths. Let's see, this first bar is one decimeter long. See, I have written 1 dm here. The next bar is one-half a decimeter long. And let me tell you the lengths of the others. When you make yours, you should write the lengths on your chart too: 1/2 dm, 1/3 and 2/3 dm, 1/4 and 3/4 dm, 1/5, 2/5, 3/5, and 4/5 dm, and 1/6 and 5/6 dm. I've made the bars representing fractions with the same denominators the same color.

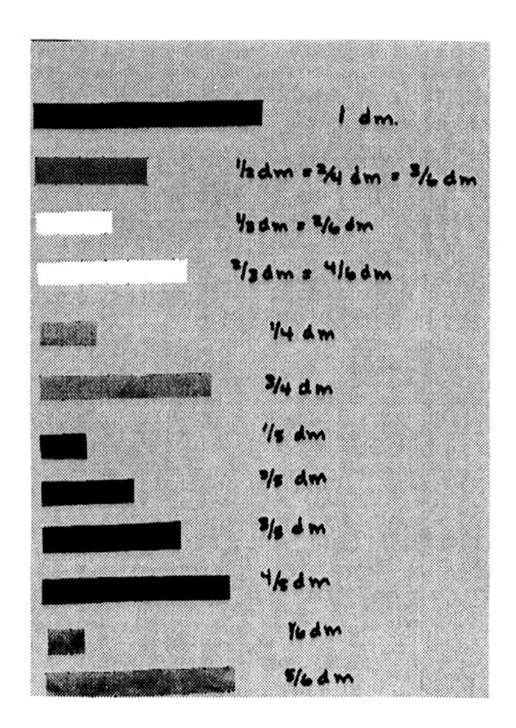

ILLUSTRATION 20. A chart showing fraction bars.

"The first bar is 1 dm long and 1 cm wide. Can you make one and cut it out and glue it to the top of your chart?"

There was some buzzing in the classroom. "What is a decimeter? Oh, yes, we learned that! It is 10 cm." The children quickly discovered that the dots on the grid paper were one centimeter apart, so making a bar 1 cm wide and 10 cm long was no problem. I worked on the overhead with a piece of transparent dot paper and colored markers. Just as they did, I cut out a 10-cm by 1-cm bar. I colored mine with a red stripe and glued it to a blank transparency. The children shared their work with each other. (Their desks were in groups of four, making communication easy.) Some were not sure if the corners of the bar needed to be on dots, or if they could just use the edge of the red paper as a boundary. (The corners do need to be on dots!) When they got the bars cut out (several tried more than once before getting an acceptable bar), they glued their bar to the top of a white sheet. "Don't forget to label it," I said. "How do we abbreviate decimeter?" They knew: it is dm. So we wrote 1 dm.

"The next bar is green, and it is one-half a decimeter long. Can you make it?" The children asked if they could use a color other than green, and I said, "Sure. The only agreement is that bars on your chart representing fractions with the same denominator be the same color." They knew that a bar of one-half decimeter would be 5 cm long; they computed this mentally. Soon each child had a 5 cm bar glued to his or her sheet. I did the same on my overhead, coloring my bar with a green marker.

"The next bar is 1/3 decimeter. How long is that?"

At this point the children got concerned. "You can't make a bar that is 1/3 dm long," said one. "There are no centimeters to match it."

"Sure you can," said another. "You could take the decimeter bar and fold it into three equal pieces. They would each be 1/3 dm long."

"Too bad we glued our 1 dm."

''Can we figure it out without cutting another one? Can we figure it out using our calculator?'' I asked.

A girl said, ''Sure, we can put 10 cm in the calculator and divide it by 3. That will tell us how many centimeters long our bar should be.'' So we did. [10][÷][3][=] gives 3.3333333 on the display.

''What is that?'' they asked.

''Gee, I don't know,'' I said. ''What do you think?''

Several were in agreement that it meant three and three-sevenths, since there were seven threes to the right of the decimal. But others weren't so sure.

''Do you think I will get 10 back if I multiply by 3?'' I asked. They thought so. We tried it: [*][3][=]. The display showed 9.9999999. The kids thought this was awesome—and that it was ''almost'' 10.

I drew a picture on the board:

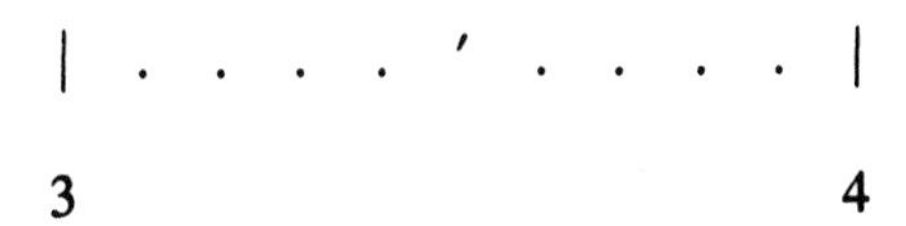

3 4

''Look at your rulers. What do you see between 3 and 4?'' Several said they saw ten millimeters—there are ten millimeters in a centimeter. The light then dawned for one boy. ''I get it!'' he said. ''You take 3 cm, and then 3 mm, and then three parts of millimeters, and then three . . .'' His voice got faint.

''How do we tell the calculator that we want 3 cm and 3 mm?'' I asked.

''3.3 cm,'' said several.

''Well, if we multiply 3.3 cm by three, will we get 10 cm?'' They tried it: [3.3][*][3][=]. The display showed 9.9.

''It is close enough,'' said one. ''I am going to make my bar 3.3 cm long.'' We agreed to this, and the children went to work.

''I chose yellow paper for this,'' I reminded them. ''If you are unsure about the measurements, you can cut yours out and check it against mine,'' I said. Many children did this; my model was passed around the class. Soon bars of approximately 1/3 dm began appearing on the children's charts. ''Be sure to label them 1/3 dm,'' I reminded them.

At this point I stopped building my transparency chart and began a large chart on the blackboard. I wrote:

1 dm = 10 cm
1/2 dm = 5 cm
1/3 dm = 3.3333333 cm

''I am going to quit making my chart on the overhead,'' I said. ''I'll just walk around in the classroom and be a helper. And I'll keep the chart on the blackboard.

''The next bar is 2/3 dm, and it should also be yellow,'' I said. I wrote 2/3 dm = on the board.

''It should be twice as long as the 1/3-dm bar,'' said some. ''But how long is that?''

''I will show you something you can do on your calculator,'' I said. ''Try [10][÷][3]—that gives us how long the 1/3-dm bar should be—and how much bigger is the 2/3-bar?''

''Two times bigger,'' chimed in several.

''So now, just continue, [*][2][=]. What do you see?''

''Six point six six six six six six six. It's like before: 6 cm and 6 mm and 6 parts of mm,'' said one.

"I am going to make my bar 6.6 cm long," announced another.

"I will make mine 6.7 cm long," said another. As before, soon bars of approximately 2/3 dm began appearing on the chart. I reminded the children to label them 2/3 dm, and I updated my blackboard chart.

"Our next bar is 1/4 dm, and I made mine orange. What do we need to do?" The children were beginning to get the procedure by this time.

"We put 10 in our calculator, and divide it by 4," said a girl. [10][÷][4][=]; the display shows 2.5.

"That is not hard," said several. "That is two and a half." Bars were measured, cut, pasted, and labeled.

"How about 2/4 dm?" I asked.

"We already have it: two-fourths is the same as one-half," said one.

"OK, how about 3/4 dm? How do we figure out how long it is?"

"It is 10 divided by 4 times 3, because we need three one-fourths," said one. [10][÷][4][*][3][=] gives 7.5 on the display. "That is seven and a half centimeters," said one.

Next we moved to fifths. "How long is 1/5 dm?" I asked.

"10 divided by 5," said one. "That is 2."

To get two-fifths, they knew the button presses were [10][÷][5][*][2][=]; the display shows 4.

"How long do you think 3/5 dm will be?" I asked. The class was already pretty confident: it will be 6 cm. But they wanted to try the program [10][÷][5][*][3][=] to be sure. Sure enough, the display showed 6.

"Four-fifths is easy," they said. "It is 8 cm."

At this point (one hour into the lesson), the fourth graders needed to go to another class. But they requested a short delay, because they wanted to know what the 1/6 dm bar looked like. "What button presses do we need?" I asked. They knew. [10][÷][6][=]; the display gives 1.6666666. By this time, deciding the length of the bar was easy.

"We will make it 1 cm and 6 or 7 mm," they said.

"Before you go, I have a challenge question for you to work on on Monday," I said. "After you finish your chart, can you put the bars in order from longest to shortest?" The kids were intrigued by this question. The longest was obvious, and they tried to estimate what would be second longest, etc. "Well," I suggested, "you can also use the chart I have made on the board here.

"Before you go, please put your name on your chart. We'll finish them Monday."

The fifth graders stayed a little longer, and we finished our charts, putting on the last bar, 5/6 dm, using the program [10][÷][6][*][5][=]; the display shows 8.3333333. But they also now needed to move to another class, so I asked if they thought they could put the bars in order, from longest to shortest. They thought they could, especially if the chart on the board was still there on Monday.

They helped me clean up and collect supplies. One girl asked me where I got the pretty paper. (They saved the pieces; they were not thrown away.) I said, "In a store."

"Why did you pick these colors?" she asked.

"Because I thought they were groovy," I replied.

"Groovy!" she said. "That's out!"

"How about 'cool'?" I asked.

"That's a little better," she answered. (This interaction reminded me of the generation gap.)

REMARKS

Many children expressed their pleasure at doing this lesson. They were very pleased with their art. Some pasted their bars, not into an orderly chart, but into patterns, still with labels. They were excited to see their calculator displays full of numbers, and they were happy that they could say what the numbers meant (3 cm and 3 mm and a little bit more). They liked it when some things they thought were not possible (e.g., dividing 10 cm into three equal parts) were actually possible. They liked seeing the threes (or the nines or the sixes) go on and on. And they liked doing the task. There was a lot of activity–computing, measuring, cutting, pasting, labeling, talking, even walking around comparing with neighbors. The papers were pretty bright colors; the glue sticks were something new. They worked together, comparing lengths and comparing art. There were lots of wonderful utterances: "Oh my gosh, look at this number!" "Did you get this?" "Wow, that is amazing." The classroom teacher told me that the fourth graders had never written fractions such as four-fifths before. But they took it in stride.

CHAPTER 24

Ants' Gardens (Grades Two and Up)

A STORY

There were two ants' nests. One was made by big Red Ants, and the other by smaller Black Ants. One day the ants decided to make gardens and plant trees in them. They also wanted to build fences around their gardens to keep grasshoppers out. The Red Ants made one big (for ants) garden, square in shape, 12 cm by 12 cm.

The Black Ants made three square gardens, one 6 cm by 6 cm, and two very small gardens, 3 cm by 3 cm each. The ants planted their trees, one in the middle of each square centimeter.

This is a make-believe story. Could real ants do it? (What do you know about real ants?) Add more to this story. What kind of trees did the ants plant? How tall was the fence? Was it wooden?

PROPS AND TOOLS

- a ruler with centimeters, a calculator, and a sheet of paper with a square centimeter grid for each child [You can xerox the grid (see Appendix). You may need extra sheets, because children will most likely make errors and want to start over.]
- an overhead calculator, a transparency of centimeter dot paper, a transparent ruler to use on the overhead, a red marker, and a black marker

THE LESSON

Children should draw the ants' nests and gardens. The teacher should draw them on an overhead transparency of centimeter dot paper. We suggest using a red marker for the Red Ants' garden and a black marker for the Black Ants' gardens. Be sure to start on a dot, and be sure that the shapes and sizes of the gardens are correct. Many younger children will make the Red Ants' garden only 11 cm long; they will begin at "1" on the ruler, instead of "0." A suggestion is to have children start at a dot, and count steps away from the dot: one, two, three. . . . Children should then draw the ants' trees. An x for a tree is enough.

How many trees did the Red Ants plant? How many trees did the Black Ants plant? How should you count them?

Children should suggest how to count the trees, but if nobody else does, the teacher should suggest the following method: for a square field, count how many trees there are in one row,

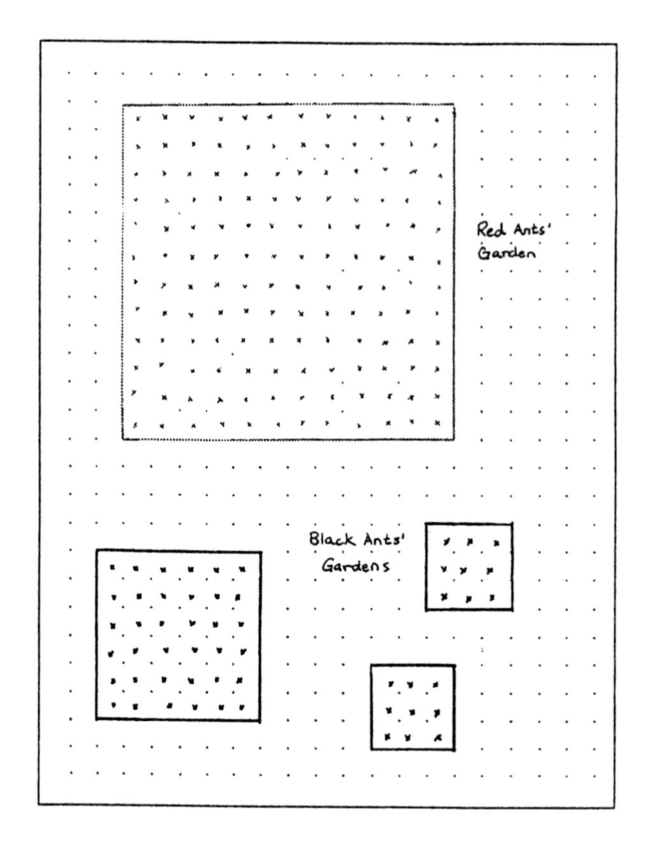

ILLUSTRATION 21. A drawing of ants' gardens.

and add this number a "proper" number of times. Example (for a 12 cm by 12 cm square garden):

[12][+][=][=][=][=][=][=][=] [=][=][=][=][=][=] Answer: 144 trees

How long was the fence around the Red Ants' garden? How long around the Black Ants' gardens? (The length should be in centimeters.)

Again, one method is to find the length of one side and add it four times. Example: [12][+][=][=][=][=]. Answer: it is 48 cm around the Red Ants' garden.

REMARKS

(1) Model programs for the number of trees in the Black Ants' gardens:

[6][+][=][=][=][=][=][=][M+] Display: 36
[ON/C][ON/C]
[3][+][=][=][=][=][=][=][M+] Display: 18
[MRC][MRC] Answer: 54 trees

Programs for length of fences around Black Ants' gardens:

[ON/C][ON/C]
[6][+][=][=][=][=][M+] Display: 24
[ON/C][ON/C]
[3][+][=][=][=][=][M+][M+] Display: 12
[MRC][MRC] Answer: 48 cm

The length of the Black Ants' fences is the same as the Red Ants' fences. The sum of the sides, 6 + 3 + 3, of the smaller squares equals 12, which is the length of a side of the Red Ants' fence. Therefore the sum of the perimeters of the small squares equals the perimeter of the big one. (This is a general fact, but the children will be surprised!)

(2) In this lesson it is more important that different methods of computations are seen than that one is chosen as "the best."

(3) This lesson is a good one for introducing multiplication.
Twelve rows of twelve trees each,

[12][*][12][=].

Four sides, six centimeter each,

[4][*][6][=]; and also [6][*]4[=].

CHAPTER 25

Geometric Progression with Squares (Grades Two and Up)

PROPS

- half a sheet of one-square-inch dot paper per child (see Appendix) (We used good quality bright red paper.)
- a ruler
- scissors
- an envelope (one per child)
- a blank sheet
- calculator
- one-square-inch dot paper for an overhead projector

THE LESSON

Task 1

Using a ruler, children should draw a square, four inches on a side, on the dot paper. They should draw lines on it at 1-in. intervals, making a 4-in. by 4-in. checkerboard. Next, they should carefully cut it out. The teacher should do this on the overhead and show the resulting square. Then they should find the area of the square (16 sq in.) and write it down on the blank page.

Task 2

Put 16 on your calculator's display: [16]. Cut the rectangle at hand into two equal (congruent) rectangles. What is the area of one of them? [÷][2][=]; the display shows 8.

Write it down on the data sheet (see below). Put one of the 8 sq in. rectangles into the envelope. (We will call it the "memory envelope.") Put the 8 on your calculator display into the calculator's memory: [M+]. Now, how much do we have in the memory envelope? 8 sq in. And how much do we have in the calculator's memory? 8.

Cut the remaining 8 sq in. rectangle into two equal (congruent) rectangles (squares). What is the area of one of them? Notice that on the calculator all you need to do is press the equals key: [=] display 4. Write it down on the data sheet (see below). Put one of the 4-square-inch pieces into the envelope. Put the 4 on your calculator's display into memory: [M+]. Now, how much do we have in the memory envelope? $8 + 4 = 12$ sq in. And how much do we have in the calculator's memory? 12.

Proceed to cut the remaining rectangles in half. Remember to put half in the envelope

ILLUSTRATION 22. Props and a finished product for geometric progression with squares.

and to press [=][M+] each time. Stop when the rectangle at hand becomes too small. (But write down its area before putting it into the envelope.) When the rectangle becomes small, the last two pieces will be the same size. Remember to put them both in the "memory envelope," and to press [M+] twice the last time through (for the two equal pieces). Now, how many square inches are in the envelope? And how much do you have in your calculator's memory?

Area of the square:
16 sq in.

Areas of parts in square inches:
8
4
2
1
0.5
0.25
0.125
0.062
0.03125
0.03125

Note:

(1) As we have done above, the areas should be computed on the calculator, using the sequence [÷][2][=][=][=] . . . , based on the principle that the area of the next rectangle is half of the previous one.

(2) The process should stop when the pieces become too small to handle, but it should be made clear that the calculation could still go on without its physical counterpart.

Task 3

In the calculator's memory you have added the areas of all the parts. Press [MRC] and you should see 16. As a check, you may add the areas of all parts which you recorded on your sheet above.

Program:

[8][+][4][+][2][+] . . . [+][.03125][=][=], or . . . [+][.03125][+][.03125][=]

Write: the sum of the areas of the pieces is 16 sq in. Notice that this sum equals the area of the original square.

Task 4

Take the pieces out of the envelope and arrange them back into a square. Put the pieces back into the envelope.

Note: for this task the pieces should not be too small. This is not a contest for the smallest piece!

Task 5

Start with a number. You may think of it as the area of a square. Divide it several times by 2. Add the parts, adding the last one twice. You get the original number.

Example:

Number:
6

Parts:
3
1.5
0.75
0.75

The sum:
6

Task 6

Here is the program we used above
(clear the calculator):

[number][÷][2]	
[=][M+]	making halves and adding them together,
[=][M+]	
[=][M+]	
[=][M+]	
[M+]	adding the last number again,
[MRC]	the display shows the original number.

Test the program.

CHAPTER 26

Square Inches

Children should make five squares from construction paper, having areas of 1, 2, 3, 4, and 5 sq in. On each square its size should be written (e.g., 3 sq in.). Each child should have an envelope to keep the squares in.

The squares should be made very precisely; the angles should be right angles, and the sides should be within 1/16 in. of the true values. The tools and techniques of drawing and cutting out squares are up to the teacher. Children should be able to read the labels (e.g., 3 sq in. should be read, "three square inches").

During this lesson children should learn how to convert decimal fractions, shown on the calculator display, into common fractions with base sixteen (sixteenths of an inch).

Each child should make at least one set of five squares, but making more copies is acceptable. The teacher should prepare his or her set beforehand, in order to be able to check quickly whether the children's squares satisfy the specifications.

Using different colored paper for each square is recommended. When this lesson was taught in a fourth-grade class, children were given five small sheets of paper, each about one-quarter of an 8-1/2-in. by 11-in. sheet, and each a different color. The sheets contained a 1-in. grid of dots, and this turned out to be a key element. Children also had a ruler, scissors, pencil, and calculator. The teacher had an overhead calculator.

THE LESSON

(1) Each child makes one square inch. It is a square whose sides are 1-in. long. (When the lesson has been taught, we have asked all children to use the same color paper for the one square inch, all to use the same color for the square with area two square inches, etc.)

(2) Each child makes a square 2-in. by 2-in. Using 1 sq in., they check that this square has an area of 4 sq in., and write 4 sq in. on it.

(3) The following question is presented, "How do you make a square that has an area of 2 sq in.?" After a discussion, the following solution should be used by all children. Make a copy of the square that is 2 in. by 2 in. Cut it into four right triangles (by drawing the two diagonals and cutting along them). Two such triangles form a square which has half of the original area. So each has an area of 2 sq in. Make a solid square of the same size. (One method for doing this is to trace around the two triangles.)

(4) The side of the last square is measured. The answer (written on the blackboard) should be the length of a side is between 1-3/8 and 1-7/16 in. (or between 1-6/16 and 1-7/16

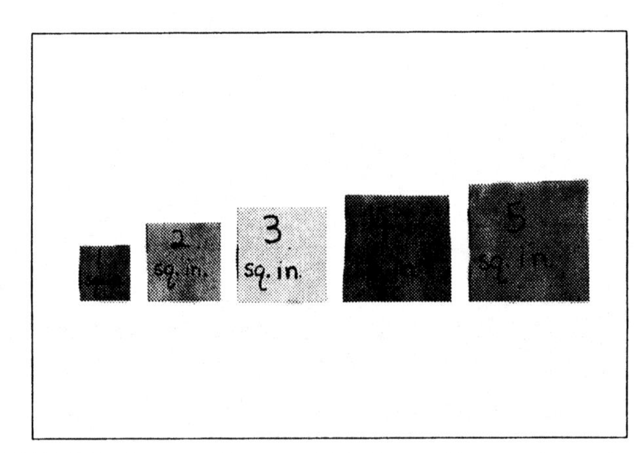

ILLUSTRATION 23. Squares with areas of 1, 2, 3, 4, and 5 square inches (not to scale).

in.). The following information should be given: If you enter the area of a square into your calculator, pressing the square root key computes the length of the side of the square.

Examples:

- The area is 1 sq in.
  [1][√] Display: 1
  The length of the side is 1 in.
- The area is 4 sq in.
  [4][√] Display: 2
  The length of the side is 2 in.
- The area is 2 sq in.
  [2][√] Display: 1.4142135

Can we say, "The length of the side is 1.4142135 in."? Yes! 1.4142135 = 1 + .4142135, and .4124235 is the calculator display of a number between 6/16 and 7/16. How can we check it? For any number n, $n = (n \times 16)/16$, so it is enough to multiply .4142135 by 16 in order to see how many sixteenths of an inch are needed.

Program:	Display:
[2][√]	1.4142135
[−][1][=]	0.4142135
[*][16][=]	6.627416

Therefore, the side of our square is approximately 1 + 6.6/16 in. long, which is between 1-6/16 and 1-7/16 in.

(5) Make a square having an area of 3 sq in. Compute the length of a side:

Keystrokes:	Display:
[3][√]	1.7320508
[−][1][=]	0.7320508
[*][16][=]	11.712812

which is a little bit less than 12. So the length of a side should be a little less than 1-3/4 inches. Make the square.

Note: this plan of action is not a recipe to be given to children. If they have been following the lesson thus far, they can make a plan themselves. But it is really less obvious than it may seem to an adult.

(6) Make a square having an area of 5 sq in.

Keystrokes:	Display:
[5][√]	2.2360679
[−][2][=]	0.2360679
[*][16][=]	3.7770864

The length of a side is approximately 2-1/4 in.

REMARKS

If this lesson takes more than one period and children are still interested, they can make more squares (not in any special order).

As children make their squares, they can stack them up, lining up a corner and two sides. In this way, the amount that is "exposed" under a square is always 1 sq in. Children will observe that this exposed shape, which looks like a capital L, gets longer and skinnier as more squares are added.

Be sure to give children an envelope to keep their squares in!

CHAPTER 27

Wipeout

Have the children enter a number on the calculator. To make the game simpler, you can have a rule that each digit in the number must be different (so 44 is not allowed, for example). For kindergartners, it can be a two- or three-digit number. For older kids it can be bigger.

Let's say the number is 7843. Now challenge them to "wipe out" the 4 (change it to zero), without changing any of the other digits. To accomplish this, the children need to subtract forty, since 7843 is 7000 + 800 + 40 + 3.

[7843][−][40][=] Display: 7803.

Wipe out the 8:

[−][800][=] Display: 7003.

Wipe out the 7:

[−][7000][=] Display: 3.

Wipe out the 3:

[−][3][=] Display: 0.

Vary the examples. Have children enter the number 87654321 and wipe it out, one digit at a time!

Now have the kids enter a number with a decimal, such as 68.43. This is trickier: 68.43 = 60 + 8 + .4 + .03. Children who have learned about money on the calculator probably know that if we are talking about money, the 4 means the number of dimes, and the 3 means the number of pennies. So how would we wipe out the 4? Four dimes make forty cents, so we have to subtract forty cents, or .40:

[68.43][−][.40][=] Display: 68.03

Wipe out the 8:

[−][8][=] Display: 60.03

Wipe out the 3:

Here again, in the context of money, the 3 means three pennies, and a penny is shown on the calculator as .01. So we have to subtract .03.

[−][.03][=] Display: 60.

Wipe out the 6:

[−][60][=]   Display: 0.

Vary the examples. Let children think of new numbers and work in small groups. One child can name the digit, and another can do the math. Try a number such as 3414. How to wipe out both of the fours, in one move? 3414 = 3000 + 400 + 10 + 4.

[3414][−][404][=]   Display: 3010.

CHAPTER 28

Target Practice

This is a game. A current display on the calculator should be changed into a specified new display (a target).

The fewer the keystrokes, the better the solution!

In general, the teacher should specify which keys should be used.

Some examples:

(1) Current display: 25
Target display: 5
Only red keys are allowed.
Some solutions:

[√]	One keystroke (Note: [√] indicates the square root key.)
[÷][25][+][1][=][=][=][=]	Nine keystrokes

(2) Current: 10.
Target: 10,000.
You must use [*] and [=].
Some solutions:

[*][1][0][=][=][=]	Six keystrokes
[*][=][=][=]	Four keystrokes
[*][=][*][=]	Four keystrokes

(3) Current: 1000
Target: 999
Any key may be used.

[=][9][9][9]	Four keystrokes
[−][1][=]	Three keystrokes

Here is one purpose of this game. There are two aspects of operations performed on a calculator. One is mathematical, and the other is purely linguistic. The mathematical aspect deals with numbers, and the linguistic with characters that you see on the display.

"Press [1][÷][2][=]. What did you get?"

"one-half" (mathematical)
"five-tenths" (mathematical)
"zero dot five" (linguistic)

Children can learn the linguistic aspect of operations before they understand their mathematical meaning.

Knowing that pressing the [+/−] key puts and removes the "−" sign in front of the display does not have to be related to negative numbers (or any numbers at all) in the minds of the children.

Similarly, knowing that if there is a "−" in front, and that then pressing the [√] key puts *E* in front does not require any knowledge of square roots.

We think that in kindergarten children can learn addition and subtraction (with a casual introduction of negative numbers: "If you subtract a bigger number from a smaller one, you get a negative number") as mathematical operations on numbers. Other operations should be used within the target game without any systematic exploration of their mathematical meaning.

The target game can be also used to illustrate some purely mathematical concepts.

For example, in how many ways can you get 3 by adding two numbers?

[1][+][2][=] [2][+][1][=]	Mathematical concept: commutativity
[3][+][0][=] [0][+][3][=]	Role of zero
[1.5][+][1.5][=]	There are other numbers besides integers.
[4][+][1][+/−][=]	You can get 3 by adding negative 1 to 4.

EXAMPLES

(1) Keys to be used: [*], [÷], [=], [ON/C], and white keys.
Problems: to achieve a target in many different ways.

- Current: 1
  Target: 1000
  Some solutions:

  [=][1][0][0][0]
  [ON/C][1][0][0][0] These are solutions which are always possible.
  [*][10] [*][10] [*][10] [=]
  [*][10] [*][=][=]
  [*][1000][=]

- Current: 1000
  Target: 3000
  Some solutions:

  [*][3][=]
  [+][=][=][=] Uses [+] key
  [+][2000][=] Uses [+] key

- Current: 80,000
  Target: 10,000
  Some solutions:

  [÷][8][=]
  [÷][2][=][=][=]
  [*][.125][=]

(2) Keys to be used: [+], [−], [=], [ON/C] and white keys.
Problems: change one or two digits.

- Current: 1,230,321
  Target: 1,230,721
  A solution: [+][4][0][0][=]
- Current: 1,230,721
  Target: 1,230,321
  A solution: [−][4][0][0][=]
- Current: 1,230,321
  Target: 1,231,121
  Some solutions:

  [+][8][0][0][=]
  [+][1][0][0][0] [−][2][0][0][=]
- Current: 1,231,121
  Target: 1,230,321
  Some solutions:

  [−][8][0][0][=]
  [−][1][0][0][0] [+][2][0][0][=]

(3) Any keys may be used.
Problems: creating patterns

- Current: 12
  Pattern: 14, 16, 18, 20, . . .
  A solution: [+][2][=][=] . . .
- Current: 8
  Pattern: 6, 4, 2, 0, −2, . . .
  A solution: [−][2][=][=] . . .
- Current: 1
  Pattern: −1, 1, −1, . . .
  A solution: [÷][1][+/−][=][=] . . .
- Current: −1
  Pattern: 1, −1, 1, −1, . . .
  A solution: [*][1][+/−][=][=] . . .
- Current: 2.0
  Pattern: 2.1, 2.2, 2.3, 2.4, . . .
  A solution: [+][.1][=][=] . . .
  Check what happen after ten presses!

(4) Any keys may be used.
Problems: undo what you did previously.

- Current: 123,456
  Press: [+][2][4][0][=] Display: 123696
  Solutions:
  [−][2][4][0][=]
  [−][=][+/−] This is a solution which has no mathematical merit. It uses specific featúres of this calculator. (It is a good example of ''hacking.'')
- Current: 123,456
  Press: [*][2][=] Display: 246912
  A solution: [÷][2][=]

- Current: 12
  Press: [*][=] Display: 144
  A solution: [√]

  Comment: be careful with the reverse order.
- Current: 12
  Press: [√] Display: 3.4641016
  A solution: [*][=] Display: 11.999999

  You did not get 12 back because of rounding of the square root and truncation of the product. (This is a fault of the calculator.)

CHAPTER 29

Patterns

"Mathematics is the science of patterns." You may have read such a statement relatively often. Observing and recognizing sequential patterns is included in mathematics curricula, often as a topic in pre-algebra.

Although we don't agree that math is the science of patterns, we think that repeated patterns of actions occur so often in all computations that they deserve special consideration.

Some examples are

- Long addition is based on the repeated pattern: add two digits and the "carry"; write one digit of the result, remember the new carry.
- Long subtraction (with regrouping): subtract the lower digit from the upper one. If the lower one is larger, regroup and either remember or write down what you regrouped. Write down the result of subtraction.
- Long multiplication requires an embedding of patterns: multiply the first number by a one-digit number and write down the result. Repeat. After you have finished, add the resulting numbers. Here, multiplying a number by a one-digit number is done by repeating: multiply two one digit-numbers, add the carry, write one digit, and remember the new carry.
- Long division is even more complex.

Programs that are presented in many lessons often give additional examples of repeated patterns of actions.

We show here a method for explaining the principles of long addition, subtraction, and multiplication with the help of a calculator.

Addition of positive numbers:

```
  1,324,567
 +9,085,072
 ----------
```

Program:

[1324567]		Display: 1324567
[+] [2]	[=]	Display: 1324569
[+] [70]	[=]	Display: 1324639
[+] [0]	[=]	Display: 1324639

[+] [5000]	[=]	Display: 1329639
[+] [80000]	[=]	Display: 1409639
[+] [0]	[=]	Display: 1409639
[+] [9000000]	[=]	Display: 10409639

(Of course, one should start with simpler examples.)

What to pay attention to:

- the same operation performed over and over again
- the pattern of numbers added
- which digits are changed on the display

In order to see what is going on, the children should perform the computations, but the results should be written on the board, one at a time.

For example:

$$\begin{array}{r} 199.23 \\ \underline{+711.574} \end{array}$$

Program:		Display:
[199.23]		199.23
[+] [.004]	[=]	199.234
[+] [.07]	[=]	199.304
[+] [.5]	[=]	199.804
[+] [1]	[=]	200.804
[+] [10]	[=]	210.804
[+] [700]	[=]	910.804

Comment: we see that a prerequisite for long addition done by hand is memorization of addition of one-digit numbers. We think that this should be achieved in the first grade.

Subtraction of positive numbers:

$$\begin{array}{r} 10{,}356.7 \\ \underline{-3618.4} \end{array}$$

Program:		Display:
[10356.7]		10356.7
[−] [.4]	[=]	10356.3
[−] [8]	[=]	10348.3
[−] [10]	[=]	10338.3
[−] [600]	[=]	9738.3
[−] [3000]	[=]	6738.3

Another example:

$$\begin{array}{r} 125 \\ \underline{-162} \end{array}$$

Program:		Display:
[125]		125.
[−] [2]	[=]	123
[−] [60]	[=]	63
[−] [100]	[=]	−37

When you subtract a bigger number from a smaller one, the result is a negative number.

Multiplication of positive numbers:

$$\begin{array}{r} 1234 \\ \underline{\times 516} \end{array}$$

Program:	Display
[1234][*][6][M+]	7404
[1234][*][10][M+]	12340
[1234][*][500][M+]	617000
[MRC]	636744

Comment: We see that a prerequisite for long multiplication done by hand is memorization of multiplication of one-digit numbers. We think that this should be achieved in the second grade.

CHAPTER 30

Bacteria

Bacteria are very small. You need a microscope to see one. But they multiply very fast. A bacterium simply divides into two new ones. Each one grows a little and divides again. And so on. Some bacteria divide very often.

QUESTION

If a bacterium divides once every hour, how many bacteria will you have after twenty-four hours in a colony which starts with just one bacterium?

SOLUTION

After one hour there will be two bacteria.

After two hours there will be four bacteria, because each of the two bacteria will divide into two.

So after each hour, the number of bacteria will increase by a factor of two. In order to compute the number of bacteria after twenty-four hours, we have to multiply 1 by 2, and repeat this twenty-three more times!

Program:

(When you run your program you have to count to twenty-four.)

Press:	Display:	Count:
[2][*][1][=]	2	one
[=]	4	two
[=]	8	three
. . .	. . .	. . .
[=]	1048576	twenty
[=]	2097152	twenty-one
[=]	4194304	twenty-two
[=]	8388608	twenty-three
[=]	16777216	twenty-four

There are sixteen million seven hundred seventy-seven thousand two hundred sixteen bacteria! Actually, the number would be slightly different because some bacteria would probably die, some would multiply a little bit faster, and some a little bit slower.

We can compute the number of bacteria without pressing the [=] key so many times. After two hours any bacterium produces four new ones (two in the first hour, and each of them will divide into two in the second hour). So in every two-hour period the number of bacteria increases by a factor of four. There are twelve two-hour periods in twenty-four hours, so we have to multiply 1 by 4, and again by 4, twelve times.

Program:

Press:	Display:	Count:
[4][*][1][=]	4	one
[=]	16	two
...	...	...
[=]	1048576	ten
[=]	4194304	eleven
[=]	16777216	twelve

We get the same answer faster.

Consider a three-hour period. In three hours one bacterium divides into eight new ones. There are eight such periods in twenty-four hours. So you have to multiply 1 by 8 only eight times. (Make your own program for it.)

If you look at a four-hour period, a bacterium divides into sixteen new ones, and there are only six such periods in twenty-four hours. (Again, you can make your own program.)

What about a five-hour period? Each bacterium divides into thirty-two new ones in five hours, but there are 4.8 such periods in twenty-four hours! You cannot press the [=] key .8 times! So look at the problem differently. In twenty-four hours there are four periods of five hours each and one period of four hours. So multiply 1 by 32 four times and then by 16 once.

Program:

Press:	Display:	Count:
[32][*][1][=]	32	one
[=]	1024	two
[=]	32768	three
[=]	1048576	four
[*][16][=]	16777216	one

You still get the same answer.

CHAPTER 31

Prime Numbers

Each integer x has at least four divisors: $1, -1, x$, and $-x$. An integer x is a prime number if $x > 1$ and $1, -1, x$, and $-x$ are the only divisors of x.

Here is a list of all the prime numbers smaller than 50: 2, 3, 5, 7, 11, 13, 17, 19, 23, 29, 31, 37, 41, 43, 47. To check if an integer x (greater than 1) is prime, check that it is not divisible by any prime number smaller than or equal to the square root of x.

Examples are (using a calculator):

(1) Is 97 a prime number?

[97][M+][√] Display: 9.8488578

We have to check divisibility by 2, 3, 5, 7.

[MRC][÷][2][=]	Display: 48.5	not divisible
[MRC][÷][3][=]	Display: 32.333333	not divisible
[MRC][÷][5][=]	Display: 19.4	not divisible
[MRC][÷][7][=]	Display: 13.857142	not divisible

Answer: 97 is a prime number.

(2) Is 91 a prime number?

[91][M+][√] Display: 9.539392

We have to check divisibility by 2, 3, 5, 7.

[MRC][÷][2][=]	Display: 45.5	not divisible
[MRC][÷][3][=]	Display: 30.333333	not divisible
[MRC][÷][5][=]	Display: 18.2	not divisible
[MRC][÷][7][=]	Display: 13	divisible

Answer: 91 is not a prime number. It is divisible by 7 and 13.

Knowing the prime numbers up to 50 allows you to check any positive integer up to 2500 for primality.

ABOUT PRIME NUMBERS

Prime numbers have fascinated people for centuries. One reason is that the primes are distributed in a very irregular way among the integers. There are many problems concerning primes that are easy to ask but difficult to answer. Problems that are easy to ask are good as enrichment material in math classes. (For a general treatment of primes see the previous discussion.) We give here two famous problems.

Twin Primes

Two odd primes are twins if their difference is two. So 11 and 13 are twins, and 17 and 19 are twins. How many pairs like that are there? Infinitely many, or not? (It is not known.)

Activities (with Calculators)

(1) Find all twin primes smaller than 100.
(2) What is the largest pair you can find?
(3) Find as many pairs as you can. (This may be a contest.)

Remark: there is only one prime triplet: 3, 5, 7. The proof is amazingly easy. It is enough to notice that among three consecutive odd numbers, one must be divisible by 3. So 3 must be part of any prime triplet (because 3 is the only prime divisible by 3).

Goldbach Conjecture

In the 1930s Goldbach conjectured that every even number greater than 2 is the sum of two primes.

$4 = 2 + 2, 6 = 3 + 3, 8 = 3 + 5$, and so on.

In all cases that have been checked, the conjecture holds, but nobody has been able to prove it in general.

Some numbers can be presented as sums of two primes in more than one way. For example:

$$16 = 3 + 13 = 5 + 11$$

Activities

(1) Check the Goldbach conjecture up to 100.
(2) Find a number, as big as you can, that can be presented as the sum of two primes in more than one way. (Get help from twin primes.)

FACTORIZATION

Factorization of a natural number (representing a number as a product of primes) is in general a difficult problem, because the process of computations requires very many steps. Using a four-operation calculator you may succeed with numbers within a range of about 10,000. (The range also depends on your skill and patience.) Using a programmable calculator, you may extend the range to millions, and using a mainframe computer, to trillions. The success in each particular case depends very strongly on the number itself. Numbers with many small divisors are easy to factor, while numbers with a few large factors are difficult to factor. Factoring a difficult number is even harder than showing that a number is prime.

Basic Method

There is only one method that can be adapted for a four-operation calculator. Here are the basic steps:

(1) Prepare ahead of time a list of prime numbers. (See the previous discussion, or use a table of primes.) The list does not have to be long. If you are not willing to do more than thirty divisions, you need only thirty primes. In order to factor a number n you never need to look at primes bigger than the square root of n. So the primes up to 31 allow you to factor every number up to 961, and also many (but not all) larger numbers.

(2) You divide your number by these primes (in any order). If the quotient is not an integer, discard the prime. If the quotient is an integer, write down the prime and continue working with the quotient. Remember, if a prime is a divisor, the quotient can still be divisible by the same prime again. The process can be speeded up if you know the criteria for divisibility by 2 and 5. (You can check divisibility mentally without punching any keys.) Even if you know the criteria for divisibility by 7, 11, and so on, do not use them because they are too error-prone.

Basic Program

You keep your current number n both in the calculator's memory and on the display. When trying a new prime p, press

[÷][p][=]

(1) If the quotient on the display is not an integer (if it has some digits after the decimal point), restore n by pressing (only once)

[MRC]

and choose the next prime.

(2) If the quotient is an integer, write down p, and press

[MRC][MRC][÷][p][M+]

and try the same p again. This sequence of steps clears the memory, computes the quotient again, and stores it back as a new n.

You see that each unsuccessful trial costs you four keystrokes, so the process can be time-consuming.

If you are not sure if the prime you are going to try is smaller than the square root of n, you may press [√] (the square root key), make a comparison, and restore n by pressing [MRC].

Examples

The primes we are going to use are: 2, 3, 5, 7, 11, 13, 17, 19, 23, 29, 31.

(1) $n = 84$.

Here, we do not test mentally for divisibility by 2 and 5, and we carry out the algorithm to the very end, ignoring that we know that 7 is a prime.

Keystrokes:	Display:	Write down:
[MRC][MRC][84][M+]	84	
[÷][2][=]	42	2
[MRC][MRC][÷][2][M+]	42	
[÷][2][=]	21	2
[MRC][MRC][÷][2][M+]	21	
[÷][2][=]	10.5	
[MRC]	21	
[÷][3][=]	7	3
[MRC][MRC][÷][3][M+]	7	
[÷][5][=]	1.4	
[MRC]	7	
[÷][7][=]	1	7

Stop. The factors are 2, 2, 3, and 7.

(2) $n = 5040$.

We simplify the process by using mental computations. We start with 2 and 5 and later go to 3, 7, and so on.

Keystrokes:	Display:	Write down:
[MRC][MRC][5040]	5040	
[÷][10][=]	504	2, 5
[÷][2][=]	252	2
[=]	126	2
[=]	63	2
[M+]	63	
[÷][3][=]	21	3
[MRC][MRC][÷][3][M+]	21	
[÷][3][=]	7	3, 7

Stop. The factors are 2, 5, 2, 2, 2, 3, 3, and 7.

(3) $n = 377$

Keystrokes:	Display:	Write down:
[MRC][MRC][377][M+]	377	
[÷][3][=]	125.66666	
[MRC]	377	
[÷][7][=]	53.857142	
[MRC]	377	
[÷][11][=]	34.272727	
[MRC]	377	
[÷][13][=]	29	13, 29

Stop. The factors are 13 and 29. We found that 29 is a prime by looking at our list of primes.

Exercises with Solutions

n:	Factors:
19110	2, 3, 5, 7, 7, 13
899	29, 31

194481	3, 3, 3, 3, 7, 7, 7, 7
1369	37, 37
103	103 (It is a prime.)

If the largest prime on your list is smaller than the square root of n (for example, the largest prime is 31 and $n = 1369$), and you have exhausted your primes without finding a factor, then the correct answer is, "I do not know what the factors of this number are."

CHAPTER 32

Fibonacci Numbers

Fibonacci (born Leonardo da Pisa, 1175–1250) was a leading mathematician of the Middle Ages. He wrote a book, *Liber Abaci,* in 1202. It was a handbook explaining how to use Hindu-Arabic numerals. (At that time, the abacus was used for calculating in Italy, and Roman numerals were used for recording results.) The book contained a problem whose solution was a sequence of integers. In the 19th century Fibonacci's name was attached to the sequence.[2]

Below we show how to create the sequence that is now called the Fibonacci numbers. It is constructed as follows: Start with 1, 1. Each subsequent number is the sum of the preceding two.

So we have:

1, 1, 2,	because	$(1 + 1 = 2)$
3,	because	$(1 + 2 = 3)$
5,	because	$(2 + 3 = 5)$
8,	because	$(3 + 5 = 8)$

and so on.

Here is a program that creates the sequence:

```
[ON/C][ON/C]
[1]
REPEAT [+][=]
```

To understand how this program works we have to learn a little more about calculators. When you perform an operation on two numbers, you see only one of them on the display. The other one is stored inside the calculator.

When you clear the calculator the number 0 is stored inside.

Keystrokes:	Display:	The number stored inside:
[ON/C][ON/C]	0	0
[1]	1	0
[+][=]	1	1

[2]The book by Theoni Pappas – *The Joy of Mathematics,* San Carlos, CA: Wide World Publishing, 1991 – contains this and more information on Fibonacci.

Keystrokes:	Display:	The number stored inside:
[+][=]	2	1
[+][=]	3	2
[+][=]	5	3
. . .	. . .	. . .

When two numbers are added, their sum is shown, and the number that you saw on the display is stored inside the calculator. Thus pressing [+][=] shows you the sum of the last two numbers you have seen before.

CHAPTER 33

Computing Lengths of Diagonals with the Pythagorean Theorem

This lesson has been taught in classes ranging from second to eighth grade. In the session described here, from a second grade class, about six children from another second grade class came in to join us.

PROPS

- a picture of three rectangles (see end of chapter) for each child (The rectangles are 9 cm × 12 cm; 6 cm × 18 cm; and 2 cm × 3 cm.)
- ruler, pencil, piece of centimeter dot paper, and calculator
- an overhead calculator, a transparency of the three rectangles, and a transparent ruler

THE LESSON

In this lesson,

(1) Children learned about the square root key on the calculator.

(2) They drew some diagonals of rectangles and measured their lengths.

(3) They were shown how to find the length of the diagonal without actually measuring its length, by using a formula (the Pythagorean theorem).

(4) They used the formula for three different rectangles, and each time they checked the calculator's result by measuring the diagonals with a ruler.

(5) They were given a piece of centimeter dot paper, to draw their own rectangles and compute and measure their diagonals.

We began the lesson with a question. The instructor asked if the children knew what the key marked with the symbol √ did. They did not know, so the instructor said, "Well, let's find out." The teacher suggested we call it the mystery key.

We decided just to try using it and to record our results. The instructor said, "Let's try [25][√]." (Enter 25 and press the mystery key.) The display read 5, so we recorded on the board: 25 leads to 5. She also said, "Let me show you a way to get back to 25: try [*][=]." (This sequence squares the number on the display.) Next she suggested they try 36. We recorded 36 leads to 6. And we pressed [*][=] to get back to 36. Next we tried 100 and recorded 100 leads to 10. (Each time we returned to the original number with [*][=].) The instructor asked if anyone had any idea what the key was doing to the numbers. One child

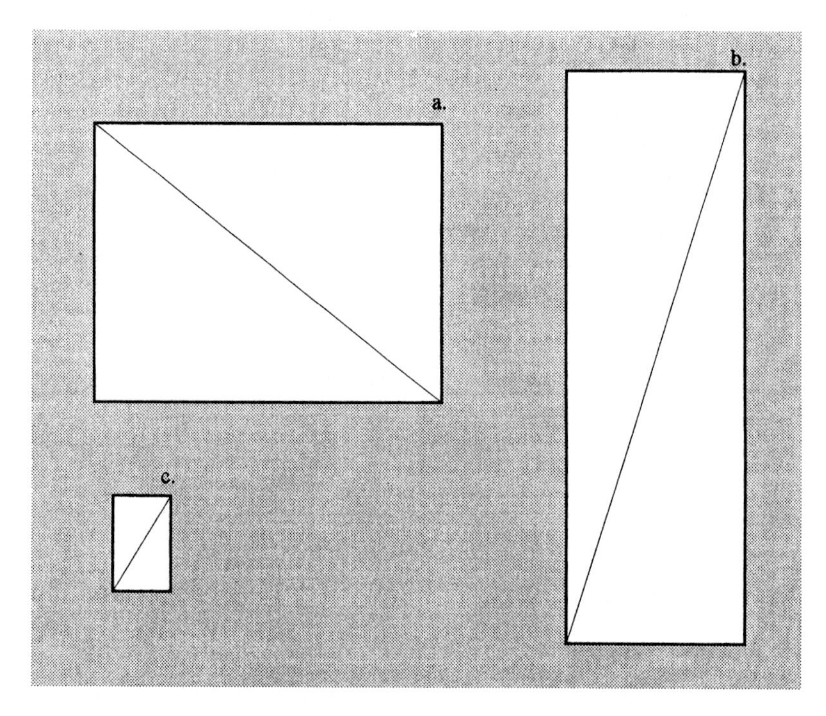

ILLUSTRATION 24. Drawing diagonals of rectangles.

said, "It takes off the tens!" (This is a clever idea and certainly holds for the first three examples!) We tried some more examples: 16 leads to 4; 9801 leads to 99; 81 leads to 9. Again the children were asked what the mystery key was doing. One child noticed that 5 times 5 is 25, another noticed that 4 times 4 is 16, and another that 10 times 10 is 100. Several children tried to state what was happening in general. By each of the numbers on the board that was gotten from the √ key, the instructor wrote, as the children observed:

25	5×5
100	10×10
16	4×4

She asked, "What will happen if you try 9 and press the mystery key?" A handful of kids knew it should be 3. We wrote $9 = 3 \times 3$. The children continued to try to verbalize what was happening. Eventually one was able to say, "You put in a number and press the mystery key, and you get a number. And when you multiply that number times that number, you get the first number back."

One child asked, "What's the key called?" The instructor wrote on the board: square root. The children pronounced it.

The instructor then said, "What do you think you'll get if you press [2][√]?" (The display shows 1.4142135, and the children were surprised.) One child read it as one point four one four two one three five.

"Now try [*][=]. Do you get 2 back?" (Actually the display shows 1.9999998.) The children were surprised by this also, but one said, "I know! It is almost 2! Like a dollar ninety nine is almost two dollars!" (The calculator truncates, so a decimal approximation to an irrational square root is smaller than the actual value. Squaring it also gives a smaller value than you started out with.)

Next we got out Picture 1 (handout at end of chapter), showing rectangles a, b, and c. The instructor put her copy on the overhead and asked the kids about the shapes. They could describe them: rectangles, with four sides and four square corners. She said, "First let's look at rectangle a. Let's use our ruler and draw a straight line from one corner to the opposite corner." She demonstrated, and the children drew. "What shapes have you made now?" The children responded that they had made two triangles that were alike. "Next question: does anybody know the name of the line we just drew, from one corner of the rectangle to the opposite corner?"

"Yes, it's a diagonal!"

She then asked them how long the diagonal of rectangle a was and how they could find out. They said they could measure and asked whether to use inches or centimeters. We decided to use centimeters. She asked that they measure and raise their hands when they thought they knew. Hands went up quickly, and there was agreement that the diagonal measured 15 cm. She said, "Write on your diagonal, '15 cm.' "

"Now I am going to show you something that is magic. Will you please measure across the bottom of your rectangle? That will tell you how wide it is. Tell me what you get. And will you tell me how tall your rectangle is?" Children measured, and agreed the rectangle was 12 cm wide and 9 cm high. "Let's label the bottom and side."

"Now here comes the magic. Try this (she wrote on the board):

[9][*][M+]
[12][*][M+]
[MRC][√]

What do you see on the display?"

Sometimes in teaching there are wonderful moments. This was one. The children were curious and mystified.

"Hey, it is 15. How did you do that?"

"What is going on here?"

"That is the length of the diagonal!"

"I want to try that again!"

The instructor suggested we try it for rectangle b. "Let's draw the diagonal. Now, what do you want to do first, measure it with your ruler, or let the calculator tell us how long it is?" They wanted to let the calculator do it. "OK, then what do we have to do?" The kids said they had to measure the bottom and the side of the rectangle. They got 6 cm and 18 cm.

The instructor erased the 9 and the 12 in the above program, leaving:

[ ][*][M+]
[ ][*][M+]
[MRC][√]

"What do we do? Let's put in our new numbers." She filled in the numbers, and read to them:

[6][*][M+]
[18][*][M+]
[MRC][√]

The display read 18.97366. She asked, "What is that? What does it mean?"

Several children said, "It is bigger than 18."

Another said, "It is almost 19."

The instructor asked the children to measure the diagonal. Lo and behold, they decided it was almost 19 cm long.

Another wonderful moment: expressions of "Cool!" "Awesome!" "Radical!!" filled the room. They wanted to do it again. So they drew the diagonal on rectangle c. They measured the sides and labeled them as 2 cm and 3 cm. The instructor erased the numbers above, again leaving:

[ ][*][M+]
[ ][*][M+]
[MRC][√]

She then asked the kids, "What do I need to fill in this time?" They were almost ahead of her, and said, "2 and 3." She filled in:

[2][*][M+]
[3][*][M+]
[MRC][√]

The display showed 3.6055512. One child said, "That is three point six and something." They measured the diagonal and found its length to be 3.6 cm.

One girl said to the instructor, "You are really smart!"

"I have to tell you something," the instructor said. "I did not figure this out. This was first figured out over 2000 years ago. Would you like to know the name of the person who first figured it out?" They definitely wanted to know. "I will write it, and you try to say it: Pythagoras. He was a Greek living in Italy. And he was pretty smart!" (Actually, according to historians, the Pythagorean theorem was known before Pythagoras!)

"So now, what have we learned here? If you know how tall a rectangle is and how wide it is, you can find out how long its diagonal is, without even measuring it! You have just used the Pythagorean theorem. And each of you has a piece of dot paper, so you can draw your own rectangles, any size you want, and measure how tall they are, and how wide, and you can use these keystrokes, which we call a program, to get the diagonal. Have fun!"

This lesson was one of the most exciting yet. The teacher said she would leave the program on the board, and they would try the dot paper in the morning.

a.

b.

c.

CHAPTER 34

Numbers in Between

I taught this lesson in a second-grade class, with the teacher present. It has been taught in grades three, four, and five as well. The concept taught was that between any two (different) numbers there is another number.

PROPS

- a metric ruler, blank sheet of paper, pencil, and calculator for each child
- an overhead calculator (preferable but not necessary)
- a blank sheet of paper tacked to the board and a black felt-tip marker

THE LESSON

"Today we are going to learn something neat about numbers. First we need to make a data sheet. Take your ruler and draw two long straight lines down your sheet, like this." I demonstrated with my felt-tip marker. "When you get them drawn, draw a line across the top, like this. See how you have made three columns? Now we will label our columns. I will label the first one 'first number.' The second one will be called 'second number.' Can you write those words? I will write them big here on the board, so you can see." I made an enlarged data sheet on the board. "Have you got it? Now, what do you think the third column will be labeled?" The kids thought it would be called "third number." "No! We will label it 'number in between.' Your data sheet should look like this:"

first number	second number	number in between

I wrote on both the paper data sheet and the board data sheet.

"OK, now we are ready for our first two numbers. Let's have our first number be 4 and our second number be 6. Can you write them on your data sheet?" I wrote:

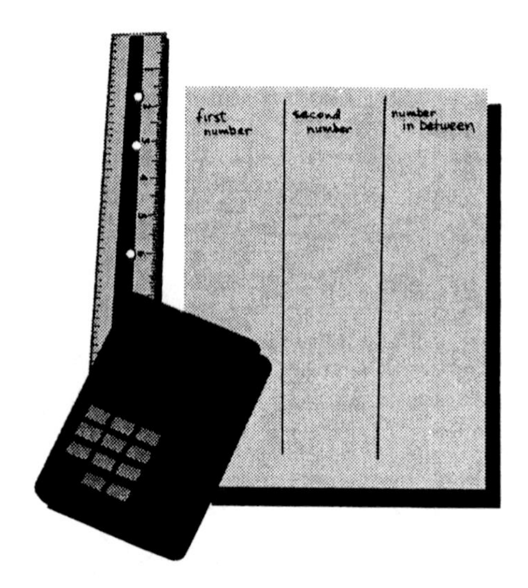

ILLUSTRATION 25. A data sheet for Numbers in between.

first number	second number	number in between
4	6	

"Can you find 4 cm and 6 cm on your ruler?" The children could do this. "Now can you find the number which is exactly in the middle between 4 cm and 6 cm?" The kids knew the number was 5. "OK, let's write 5 down in the third column, under 'number in between.'

"Here is another pair of numbers: the first number is 10, and the second is 20. Can you write them on your data sheet? Here is how my data sheet looks now:"

first number	second number	number in between
4	6	5
10	20	

"Can you find 10 cm and 20 cm on your ruler? What number is exactly in the middle between 10 and 20?" The children knew it was 15. "So write 15 on your data sheet. Let's try another pair. Write 8 as the first number and 12 as the second. Got it? Now find 8 cm and 12 cm on your ruler. What number is exactly in the middle between 8 and 12?" The children could see it was 10. "Write 10 in the third column on your data sheet.

"Let's do one more, and then we will learn something pretty exciting. Let's choose 12 as our first number and 24 as our second. Can you record them on your data sheet? It should now look like this."

first number	second number	number in between
4	6	5
10	20	15
8	12	10
12	24	

"Can you see on your ruler what number is exactly in the middle between 12 cm and 24 cm? This is getting a little tricky!'' A few children were able to get the number 18, but it was difficult.

''OK, now we will learn some magic. Try this.'' I wrote on the board:

[12][+][24][÷][2][=]

''Twelve plus twenty-four divided by two equals. What do you see on your display?'' 18. I also showed the keystrokes on the overhead calculator, and I waited until each child was able to press the correct keys and get the answer. Some children said it was magic, just like the ''magic'' program for diagonals I had shown them a few weeks before (see Chapter 33, ''Computing Lengths of Diagonals with the Pythagorean Theorem'').

''Let's check to see if this program works for the numbers in our table so far. I will erase the 12 and the 24, and you tell me what I need to do to check the first row.''

I left on the board:

[ ][+][ ][÷][2][=]

''Can you tell me what I need to fill in to check the first row?'' The kids suggested to put 4 in the first blank, and 6 in the second. I wrote:

[4][+][6][÷][2][=]

''Try it! If you get the answer that is on your data sheet, put a check mark by it, in the third column.''

This caused some excitement in the class. Some children went ahead and checked all their entries. When children got stuck, I went through the keystrokes with the overhead calculator. We checked all the number pairs in our table.

''Now let's try some new number pairs. How about 12 and 13? What is the number that is exactly halfway in between? Shall we use our program? Let me erase, and you tell me what to fill in.'' The children said,

[12][+][13][÷][2][=]

''What do you see?''

''12.5 (twelve point five).''

''What does it mean?'' Some children remembered that ''point five'' is how the calculator shows one-half. George was about to pop. He said, ''It is twelve and a half. Twelve and a half is halfway between 12 and 13.''

I continued with harder examples. ''Let's try 15 and 15.2. What will be halfway in between? Try it!''

Keystrokes: [15][+][15.2][÷][2][=] Display: 15.1

I asked, ''What does it mean?''

Andrew said, ''15.1 is halfway between 15 and 15.2.''

I asked, ''Can you find it on your ruler? Can you find 15 cm and 15.2 cm, and then 15.1 cm?'' Most children could do it.

We continued with the following number pairs:

- 11.3 and 21.4

- 12.1367 and 18.3568
- 10.0003 and 10.0004

This last pair was especially "awesome": the midpoint is 10.00035. "Can you find it on your ruler?"

"No way," said some kids. "It is too small." In every case, children read the answers: ten point oh oh oh three five."

The children also proposed a few pairs, and we used our program to find the number halfway in between.

At the end, the data sheets looked like this (some had recorded a few more pairs, suggested by children):

first number	second number	number in between
4	6	5
10	20	15
8	12	10
12	24	18
12	13	12.5
15	15.2	15.1
11.3	21.4	16.35
12.1367	18.3568	15.24675
10.0003	10.0004	10.00035

I asked the kids, "So what did you learn about numbers today? Can anyone say? If you think you know, raise your hand!"

This brought about an interesting discussion. The final consensus was, if you have two numbers (no matter how "crazy" they are), you can find the number that is exactly halfway between them.

One not especially mathematically inclined kid asked me, "Is it important to divide by 2? Or can I divide by something else and always get something in between?" I showed him, using 4 and 6, that when you divide their sum by 2, you get 5. When you divide by a number close to 2, like 1.9 ([4][+][6][÷][1.9][=]), you get 5.2631578. That is between 4 and 6. But you can't just divide by anything. For example, when you divide the sum by 5, you get 2, which is not in between.

REMARKS

If there had been more time, I would have used some pairs in which the first number was bigger than the second, to show that order doesn't matter. I also might have used examples in which one of the two numbers is zero, or one or both numbers are negative. In the lesson as it was taught, I used only numbers which could be found on a 31-cm ruler.

CHAPTER 35

How Thick Is a Sheet of Paper? (Grades Three or Four and Up)

TOOLS

- a ruler with a centimeter and millimeter scale and a calculator for each child
- a book for each group of two to four (The book can be the same for all groups, or for variety, each group can have a different book.)

THE LESSON

The question is given in the title. Let children give their ideas!

A sheet of paper is too thin to be measured by a ruler, but we can measure the thickness of a page in a book rather easily. Measure the thickness of the book (without its cover) and divide the result by the number of sheets. There are two things to watch out for here:

(1) The number of (numbered) pages in a book is twice the number of (numbered) sheets of paper in the book.
(2) There may be an "extra" page or two at the beginning and end of the book.

So look in the book. If all the pages are numbered, beginning at page 1, then if the number of (numbered) pages is even, the number of (numbered) sheets of paper is the number of pages divided by 2. Often the "extra" pages are thicker than the others. You may choose to omit them from your measurement just like you omit the cover. Does this make a difference in your answer?

Example:

Thickness of the book: 2.1 cm. Number of pages: 526.

[526][÷][2][=] Display: 263

There are 263 sheets of paper in the book.

[2.1][÷][263][=] Display: 0.0079847

Answer: one sheet of paper is 0.008 cm (0.08 mm) thick.

Do some books have thinner pages than others?

We have taught this lesson a number of times. In one version, we gave the children old elementary school math books from the nineteenth and early twentieth centuries. We recorded on the board the copyright date of the book and the thickness of a page from the book. Children were able to determine whether older books have thicker pages.

ILLUSTRATION 26. Finding the thickness of a page in a book.

CHAPTER 36

Words, Words, Words . . .

PROBLEM

How many words are in a book? (Choose any book that children have read recently.)

Method

Different methods of finding the number of words should be discussed in detail and compared. Here are some that can be considered:

(1) Count them. Good point: you get the exact answer. Bad point: too much work. Maybe teamwork would help? Children could count words on different pages. How many pages per child? What about errors in counting?
(2) Count words on a page and multiply by the number of pages. Good point: less work. Bad point: it gives only an approximation; different pages have different numbers of words. Possibility of improvement: compute the average from a few pages. Question: how to choose these pages?
(3) A method that is simpler still: choose some pages. Count the number of lines on each page. Compute the average number of lines per page. Choose some lines and compute the average number of words per line. Multiply the average number of words per line by the average number of lines per page by the number of pages.

Activity

Divide the class into groups of three to six children. All groups get identical copies of the same book (so comparisons can be made later). Each group can choose its method. Even if all choose the third method, each group should choose its own pages and lines to use. Each group does its counting and computing and presents its estimate of the number of words.

Final Discussion

Results are compared and methods (especially the choice of lines and pages) are discussed.

Note: Choosing pages and lines at random is a possibility, but not necessarily the best one. Without estimating the size of the sample needed to get a reliable result (which is too difficult for children), a random choice cannot be justified. Looking through the book and choosing "typical pages" and "typical lines" and taking into account the beginning of chapters, pictures, and so on, in a systematic, nonrandom manner, can give a better estimate.

Additional Questions

How fast do you read? How many pages per hour? Words per minute? Does it vary? Do you remember the content better when you read slowly? How long do you think it would take you to read the book above?

CHAPTER 37

Computing Areas

PROPS AND TOOLS

- rulers
- (convex) polygons drawn on square 1-cm grid paper with vertices located on points of the grid (see first handout at end of chapter), one for each child

THE LESSON

This lesson shows children how to compute the area of the polygon in square centimeters. Children should know the formula for the area of a rectangle:

area = base × height

and for the area of a triangle:

area = base × height/2

Method

(1) Draw and label the polygon. Circumscribe a (big) rectangle around the polygon. Divide the part of the rectangle lying outside the polygon into triangles and (small) rectangles. Do this by drawing horizontal and vertical lines only. Label the triangles and (small) rectangles (see second handout).

(2) Measure, record measurements, and compute areas. Children should measure heights and bases of the triangles and rectangles by counting the number of 1-cm segments (measuring with rulers is optional). The areas can be computed mentally or with calculators. Children should prepare their own tables of recorded measurements. (Do not prepare worksheets for this purpose.) An example of recorded data is given below:

	base:	height:	area:
Big rectangle	15	20	300 sq cm
Small rectangles			
A	5	3	15

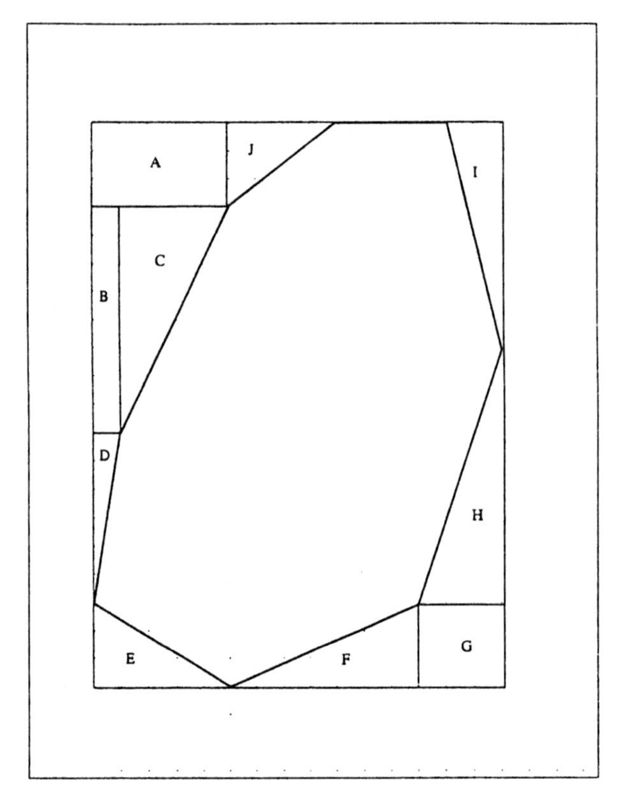

ILLUSTRATION 27. Computing areas.

	base:	height:	area:
Small rectangles			
B	1	8	8
G	3	3	9
Sum of the areas			32 sq cm
Triangles			
C	4	8	16
D	1	6	3
E	5	3	7.5
F	7	3	10.5
H	3	9	13.5
I	2	8	8
J	4	3	6
Sum of the areas			64.5 sq cm

(3) Make the final computation. The area of the polygon is equal to the area of the big rectangle minus the total areas of the small rectangles and the total area of the triangles. The area of the polygon is $300 - 32 - 64.5 = 203.5$ sq cm. (We can measure this area in different units, getting 20350 sq mm, 2.035 sq dm, and 32.56 sq in.)

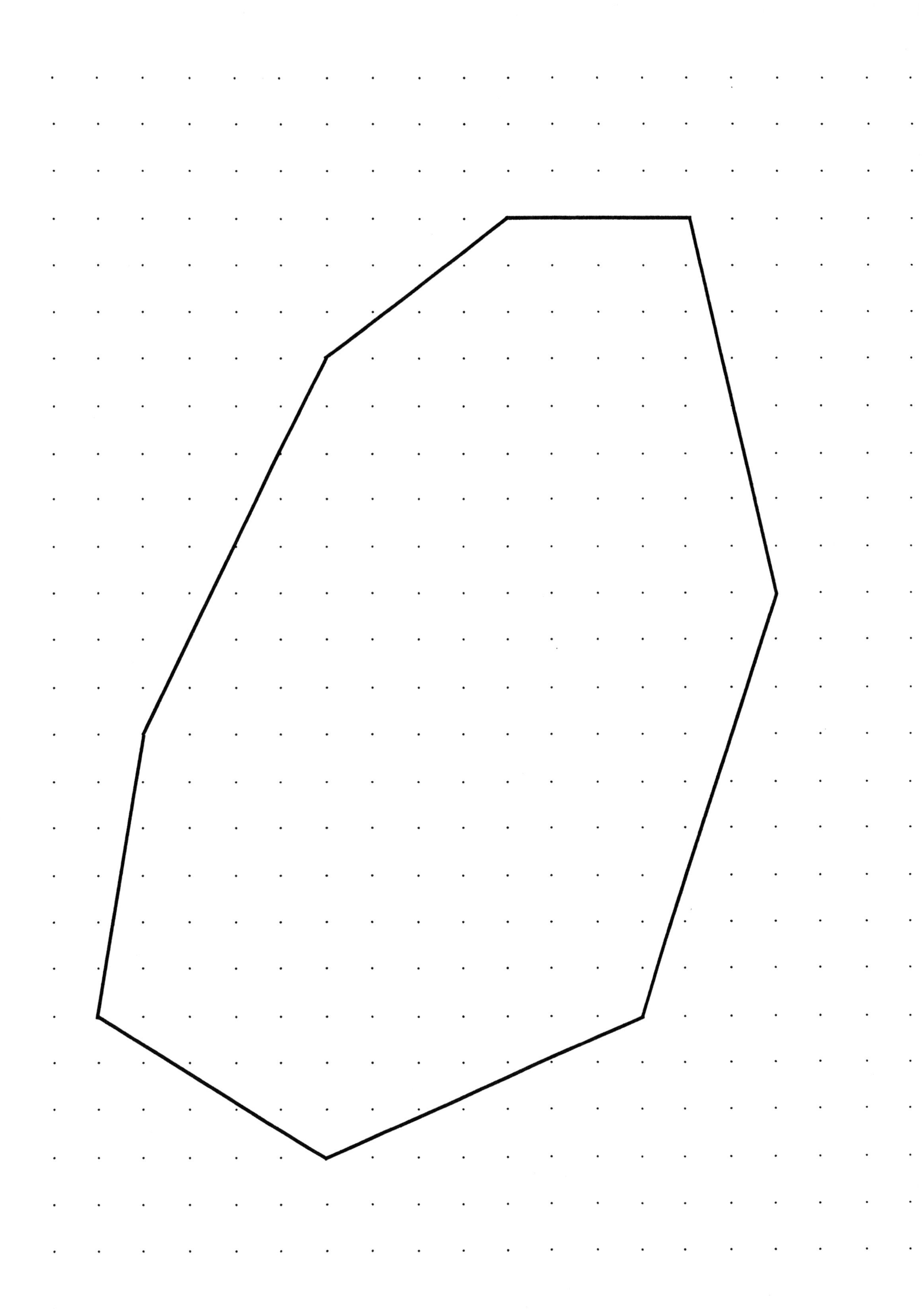

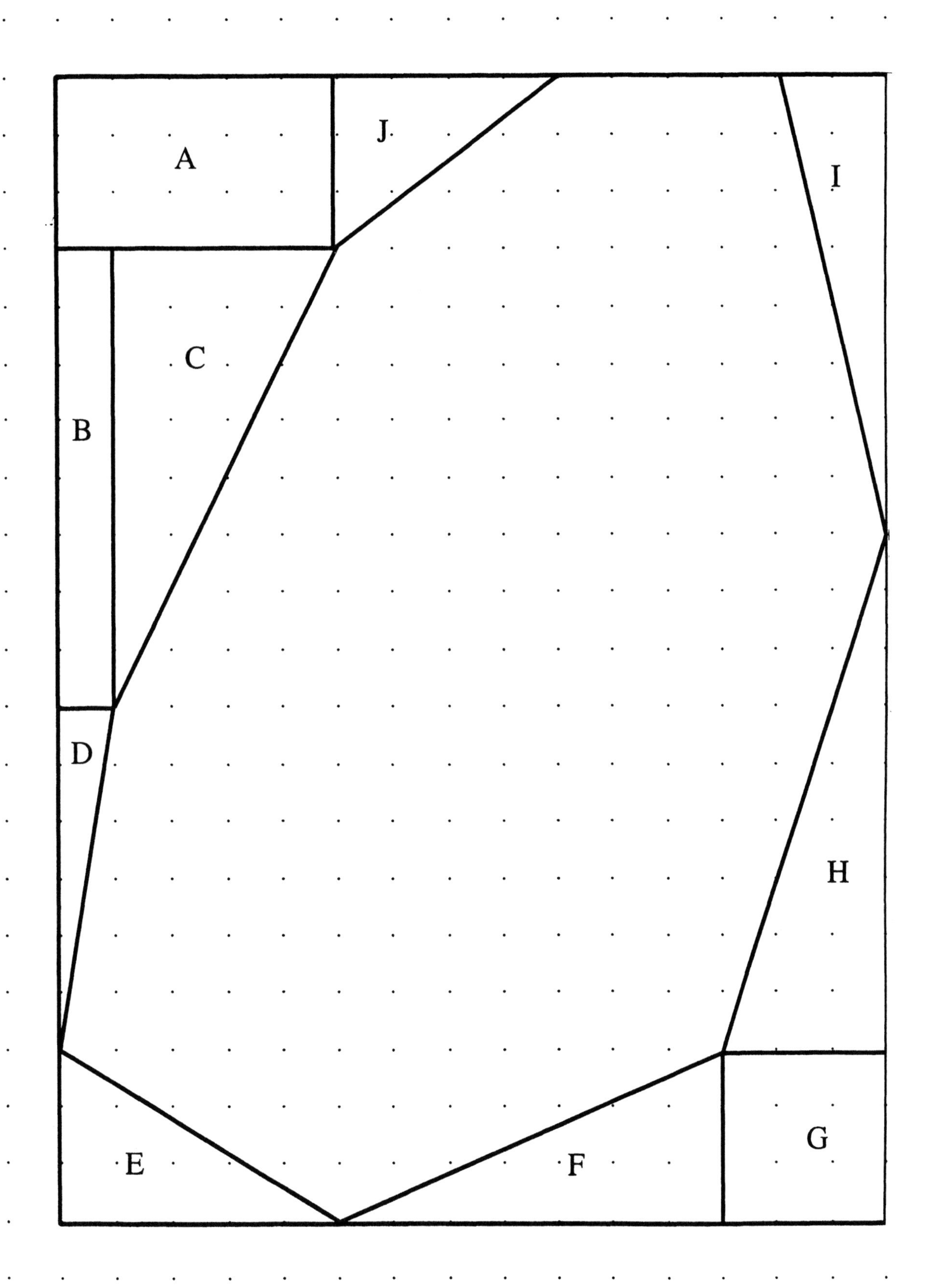
A
J
I
C
B
D
H
G
E
F

CHAPTER 38

Area of a Triangle (Using Heron's Formula)

The area of a triangle can be computed from the formula:

area = (base × height)/2

But how can we compute the area when the lengths of the three sides a, b, and c, of the triangle are given? The following formula was discovered by Heron of Alexandria (first century A.D.):

area = sqrt($s \times (s - a) \times (s - b) \times (s - c)$)

where $s = (a + b + c)/2$, and sqrt is the square root. This formula is not convenient for hand calculations but can be computed with a calculator. The computation consists of two parts:

- computing s, $s - a$, $s - b$, and $s - c$
- computing the square root of their product

PART 1

Write a list of four lines:

First	$s =$
Second	$s - a =$
Third	$s - b =$
Fourth	$s - c =$

Compute:

[MRC][MRC][ON/C][ON/C]
[*a*][+][*b*][+][*c*][÷][2][=][M+]

Write the result in the first row.

[−][*a*][=]

Write the result in the second row.

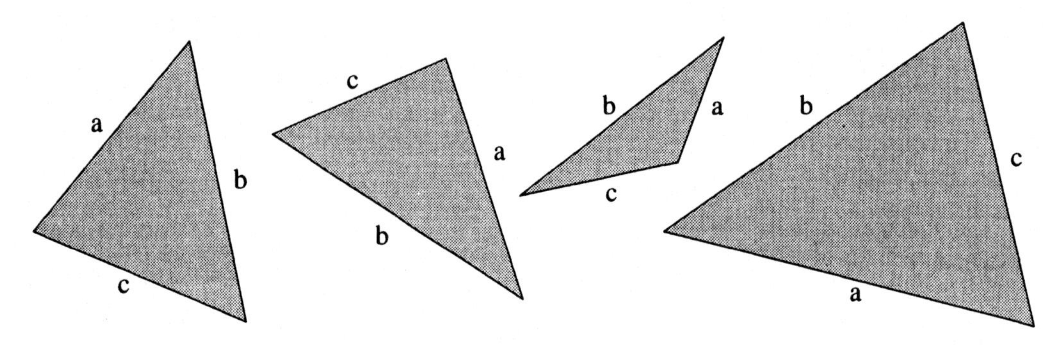

ILLUSTRATION 28. Triangles.

[MRC][−][*b*][=]

Write the result in the third row.

[MRC][−][*c*][=]

Write the result in the fourth row.

PART 2

Write: the area =

Compute:

[MRC][MRC][ON/C][ON/C]
[first row][*][second row][*][third row][*][fourth row][=][√]

Write the answer.

Example:

Compute the area of a triangle with sides $a = 1.5$ in., $b = 2$ in., and $c = 2.5$ in.

First	$s = 3$
Second	$s - a = 1.5$
Third	$s - b = 1$
Fourth	$s - c = 0.5$

The area = 1.5 sq in.

Comment: when numbers are as simple as in this problem, the first part of the computation should be done mentally.

CHAPTER 39

Area of a Polygon

Prerequisite: computing the area of a triangle by Heron's formula.

PROPS AND TOOLS

- polygons cut from stiff paper or cardboard, or even drawn on paper (see handouts at end of chapter)
- ruler with cm and mm markings

THE LESSON

(1) Divide the polygon into triangles by drawing straight lines. Mark the edges and lines by letters $a, b, \ldots$ (see Illustration 30). Notice that a polygon can be divided into triangles in many different ways. This does not influence the final result. The differences among results will be because of unavoidable errors in measurements and rounding errors during calculations.

(2) Measure all sides of the triangles, with an accuracy of one millimeter, and write down the results. The results for a figure may look as follows:

$a = 7.5$ cm
$b = 8.4$ cm
$c = 8.2$ cm
$d = 5.7$ cm
$e = 8.0$ cm

(3) Compute the areas of all triangles using Heron's formula:
Triangle 1 (sides a, b, and c):

$s = 12.05 = (a + b + c) \div 2$
$s - a = 4.55$
$s - b = 3.65$
$s - c = 3.85$
area 1 $= 27.757222 = \text{sqrt}\,(s \times (s - a) \times (s - b) \times (s - c))$

Triangle 2 (sides c, d, and e)

$s = 10.95$
$s - c = 2.75$

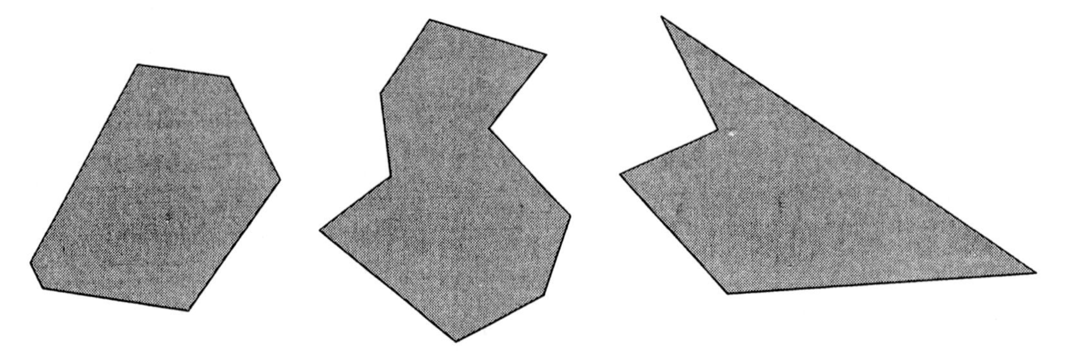

ILLUSTRATION 29. Polygons.

$s - d = 5.25$
$s - e = 2.95$
area 2 $= 21.595539$

(4) Add the areas of the triangles and round the results.

The area of the polygon = 49.35 sq cm.

(5) If you want to know how to round reasonably, you should know how the error of the computed area of one triangle depends on errors in your measurements of its sides. Even if you are very careful, the error in the measurement of each side can be half a millimeter, or 0.05 cm. If you know higher mathematics (calculus), you can derive a formula which says that the error in the area can be estimated as:

$(2 \times \text{area}/s) \times$ (error of measuring a side)

So the error for the first triangle is

$(2 \times 28/12) \times (0.05) = 0.23$ (rounded).

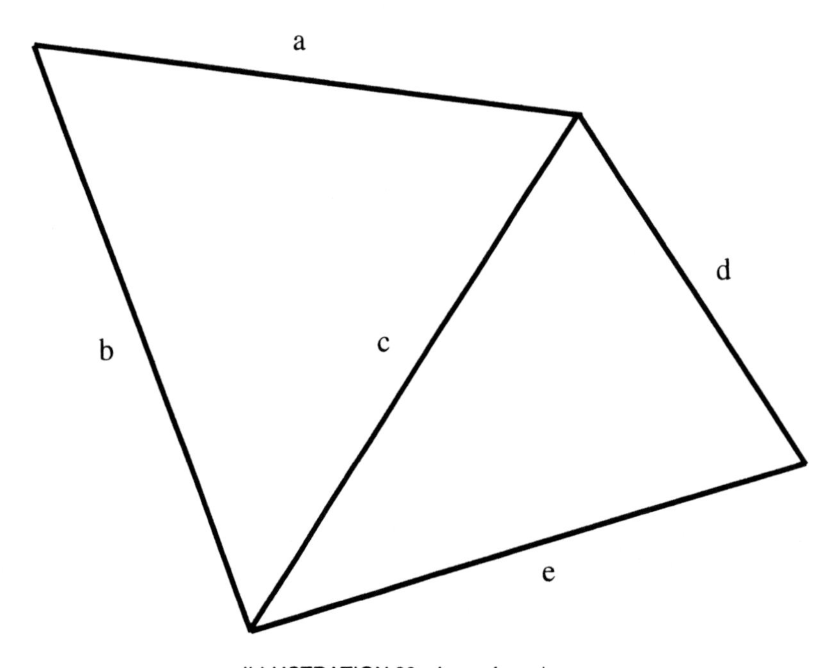

ILLUSTRATION 30. Area of a polygon.

And for the second triangle:

$(2 \times 22/11) \times (0.05) = 0.22.$

The total error may be as big as 0.45 sq cm, and we do not know whether the real value is bigger or smaller than the computed one. We are reasonably confident that the area is somewhere between 48.9 and 49.8 sq cm.

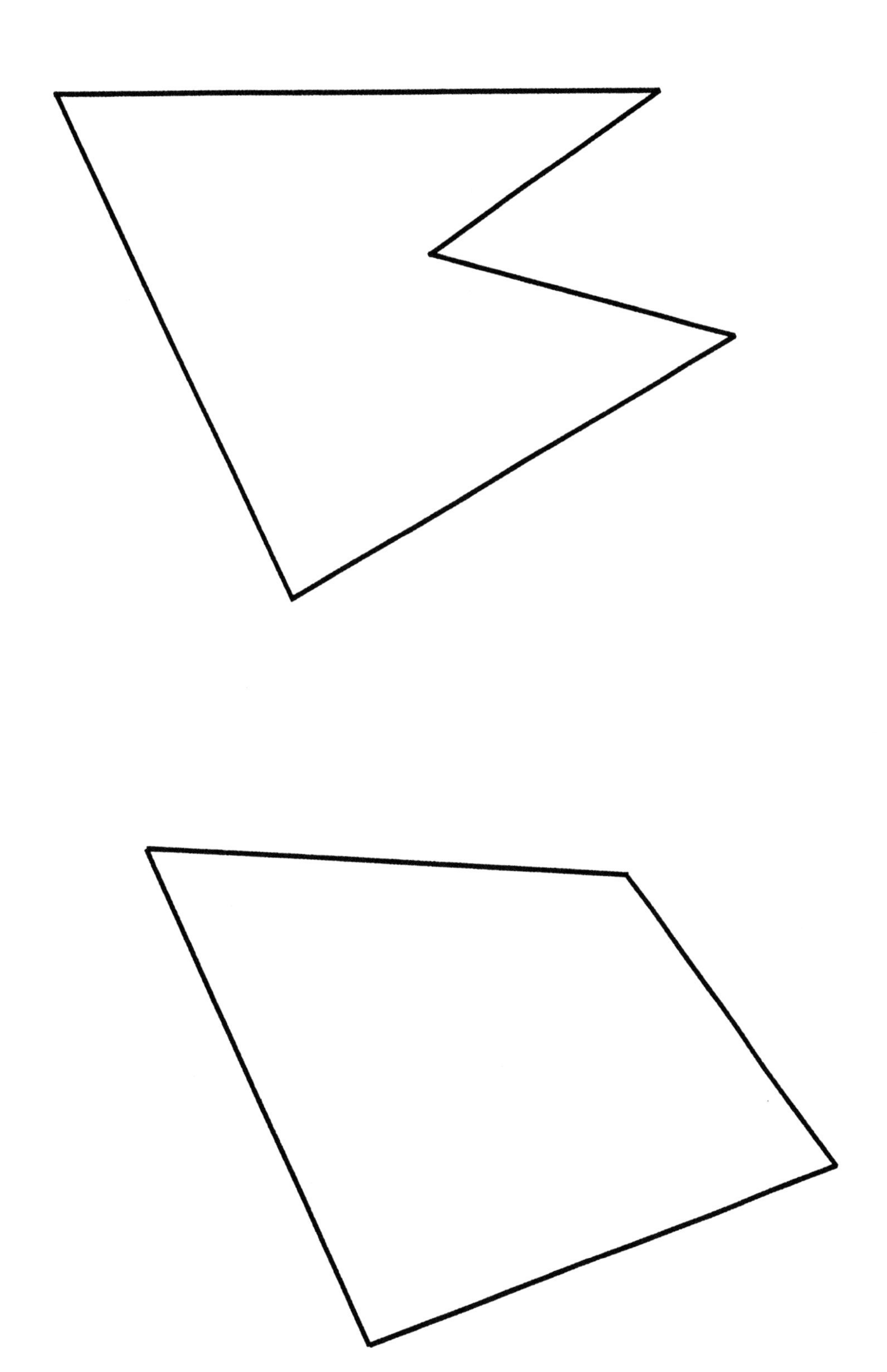

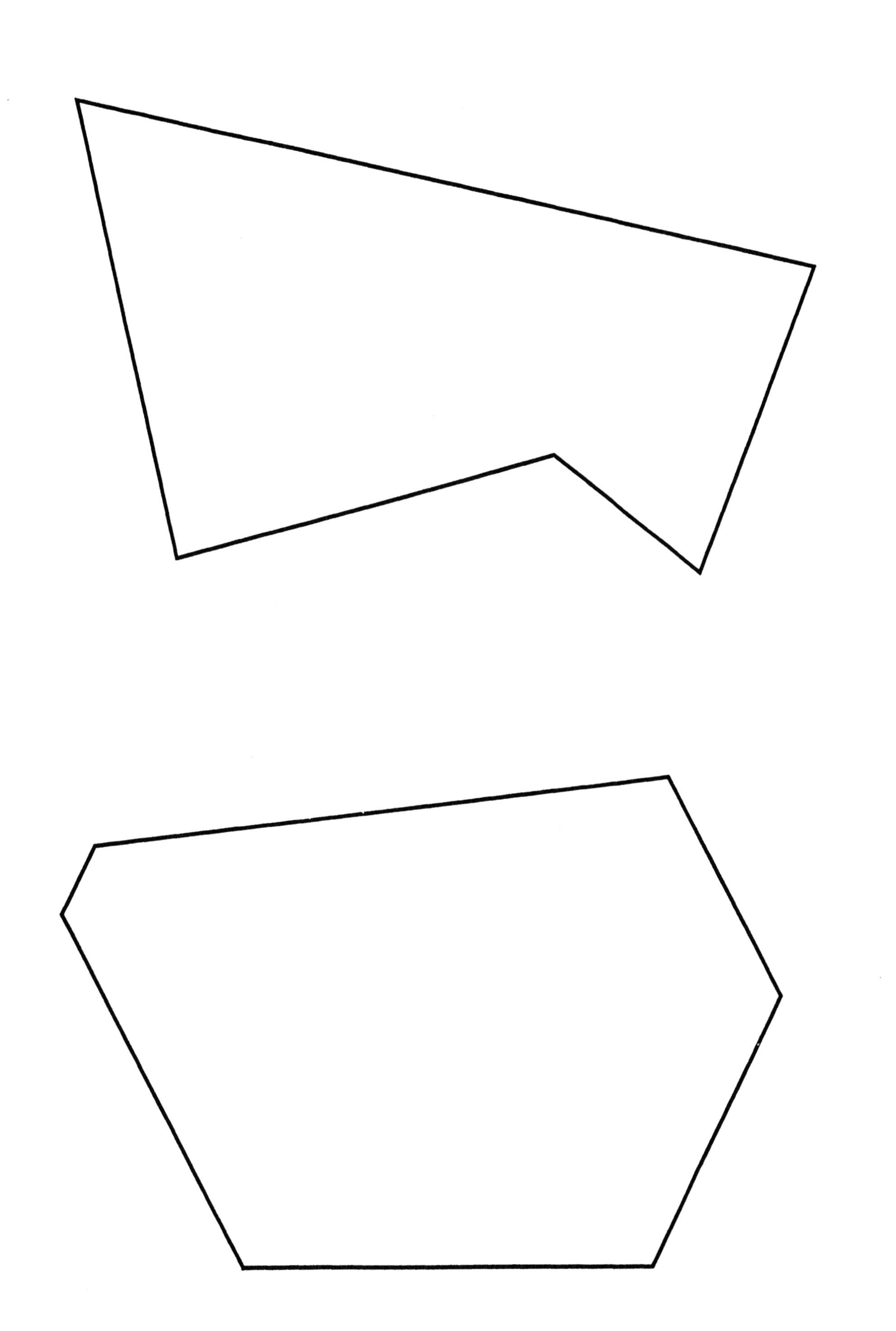

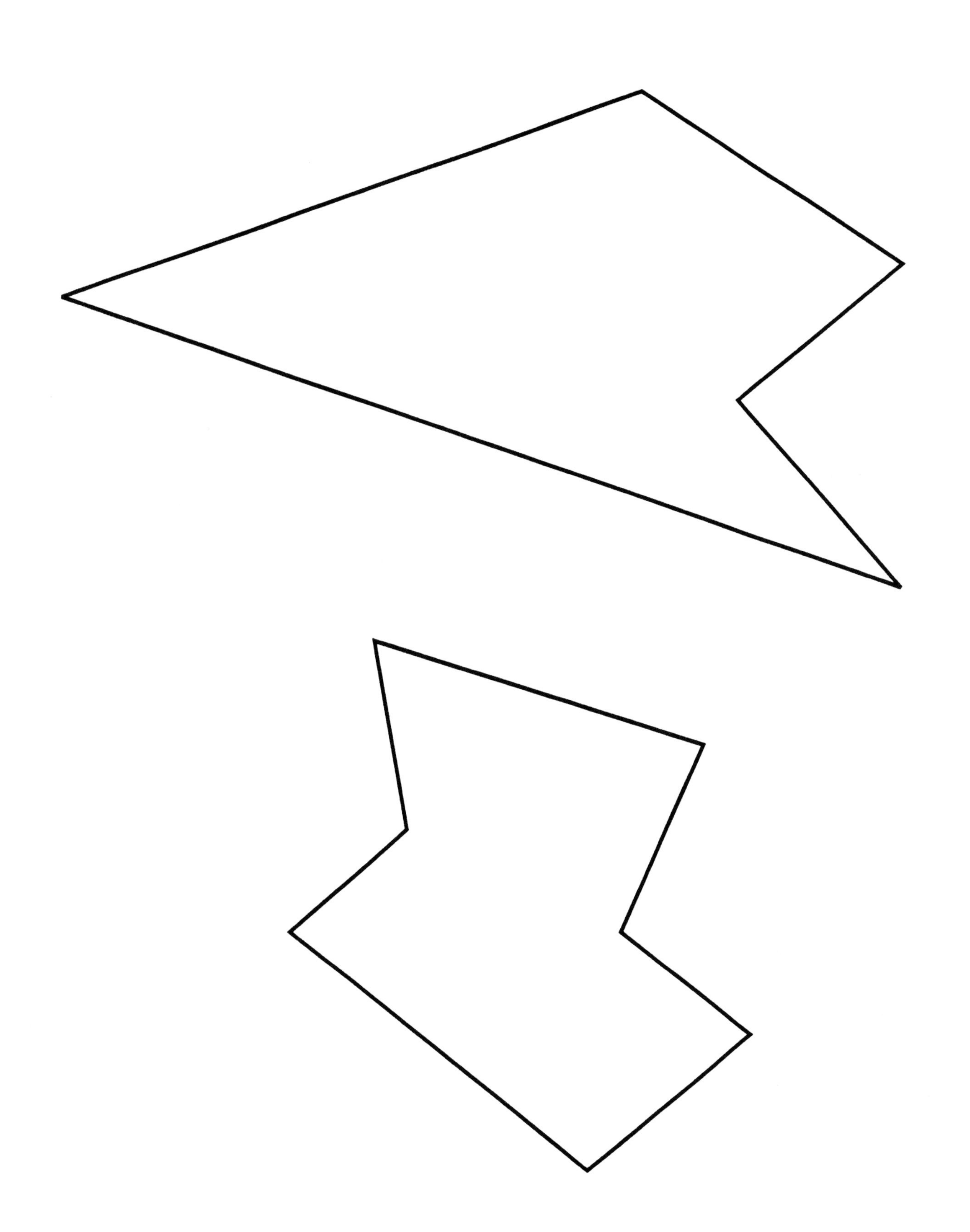

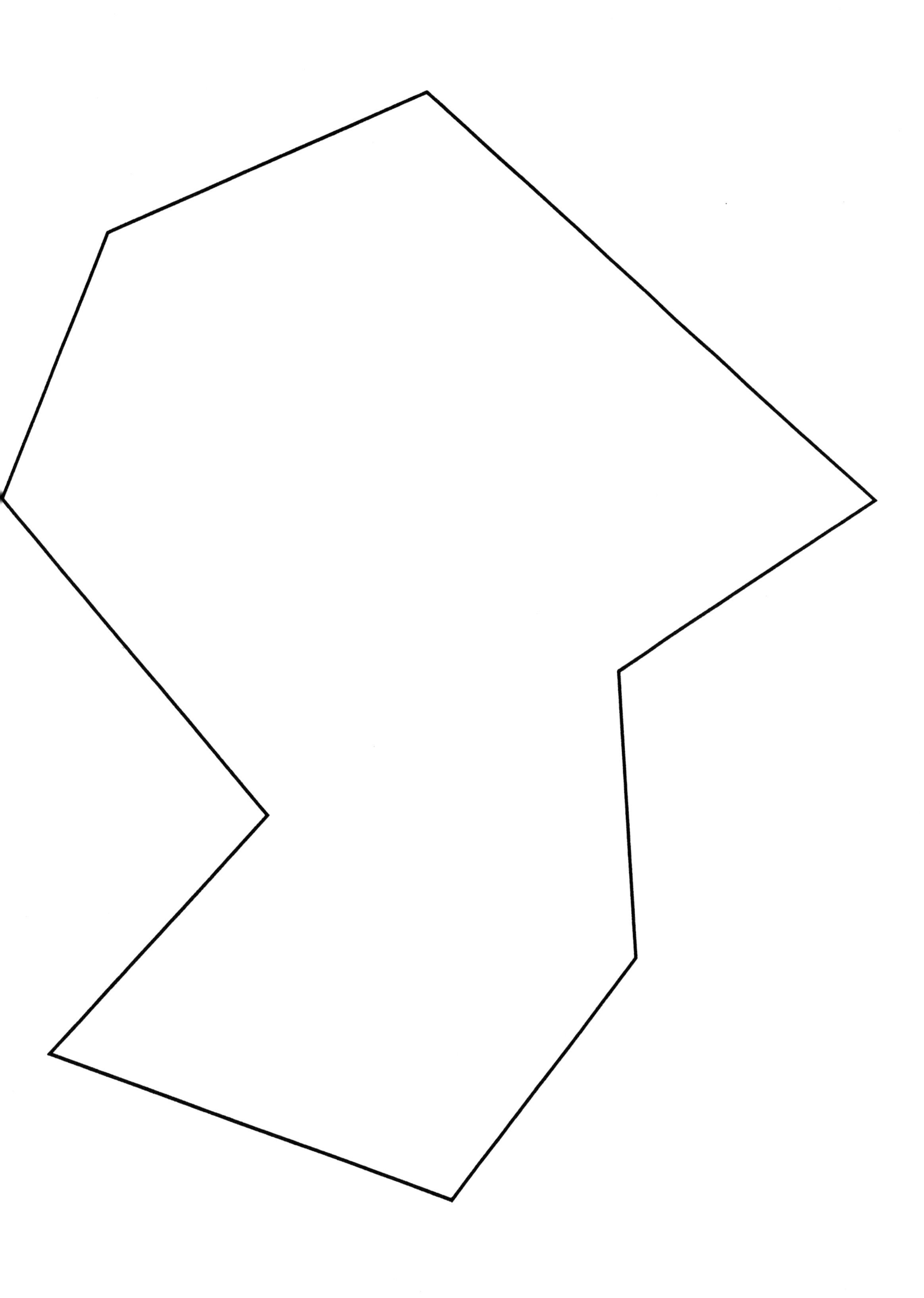

CHAPTER 40

Leaves

This lesson shows children how to measure the area of a maple or oak leaf (or other suitable leaf). It should be taught in connection with topics on photosynthesis, plant classification, and plant structure.

There are many ways of doing this lesson. Probably the most common is to draw or paint an outline of the leaf on a 1-cm square grid and estimate the area of the drawing by counting the number of whole squares covered and giving partial credit for partially covered squares. Here we present a different method.

PROPS AND TOOLS

- one sheet of 1-cm dot paper (see Appendix), ruler, calculator, paper, pencil, cellophane tape, and a leaf for each pair of children
- an overhead transparency of 1-cm dot paper, and transparent ruler for the teacher

THE LESSON

Put the leaf anywhere on the grid and fix it with transparent tape so it does not slide. Count the number of dots that are covered (are not visible). Write down the number. Repeat several times, changing the position of the leaf. If you get a value that is very different from the rest, discard it—you made an error. The average of the remaining values gives you the leaf's approximate area in square centimeters.

How do you count the invisible points? Go line by line. If you are not sure how many points are covered, measure the distance between the two closest visible points in this line and subtract one.

Example:

```
.    .    .    .    .    . / / / / / / / / / / / / / / / .    .    .    .    .
                         |--------5 cm---------|
```

Four points are covered.

Do not write down all the values. Subtract one mentally and use the calculator to keep the current sum. (Working with a partner helps!)

Why does counting points give an estimate of the area? Think of each point of the grid as

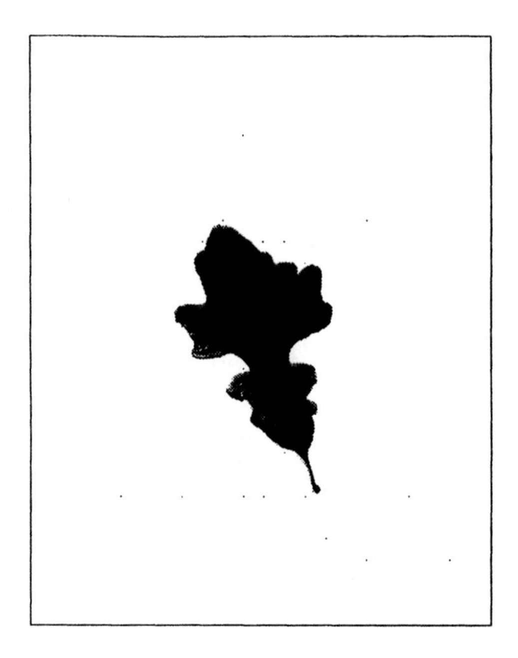

ILLUSTRATION 31. Measuring the area of an oak leaf.

the center of one square centimeter. You count all the square centimeters that are completely covered and those near the border whose center is covered by the leaf. The actual number of points covered depends on the position of the leaf on the grid. In order to remove this effect, the measurements are repeated and the average is taken.

Example (measurement of an elm leaf):

Trial:	Value:
1	52
2	46
3	64
4	49
5	50
6	46

The value from trial three (64) was discarded. The average of the remaining five is 48.6, so the area is estimated as 49 sq cm. (This whole procedure took approximately ten minutes.)

CHAPTER 41

Area of a Circle

PROPS AND TOOLS

- rulers (with centimeters)
- figures drawn on 1-cm dot grid paper (see handout at end of chapter), one sheet per child

THE LESSON

Task 1

What is the area, measured in square centimeters, of the top square (see handout)?

First method: measure the length of a side with a ruler (12 cm) and compute its square: [12][*][=], the display shows 144.

Second method: to the right of the big square there is a picture of one square centimeter with a dot in the middle. In order to know how many such small squares fit into the big one it is enough to count all the dots inside the big square. We have twelve rows, each having twelve dots. [12][*][12][=]. The answer is the same: 144 square centimeters.

Task 2

What is the area of the circle drawn inside the square below?

First method: let's count the dots inside the circle. This gives only an approximation of the area, because some of the 1-cm squares are only partway in the circle. Let's count the dots row by row, adding the numbers:

Row number:	Count:	Press:	Display:
1	4	[4][+]	4
2	8	[8][+]	12
3	10	[10][+]	22
4	10	[10][+]	32
5	12	[12][+]	44
6	12	[12][+]	56
7	12	[12][+]	68
8	12	[12][+]	80

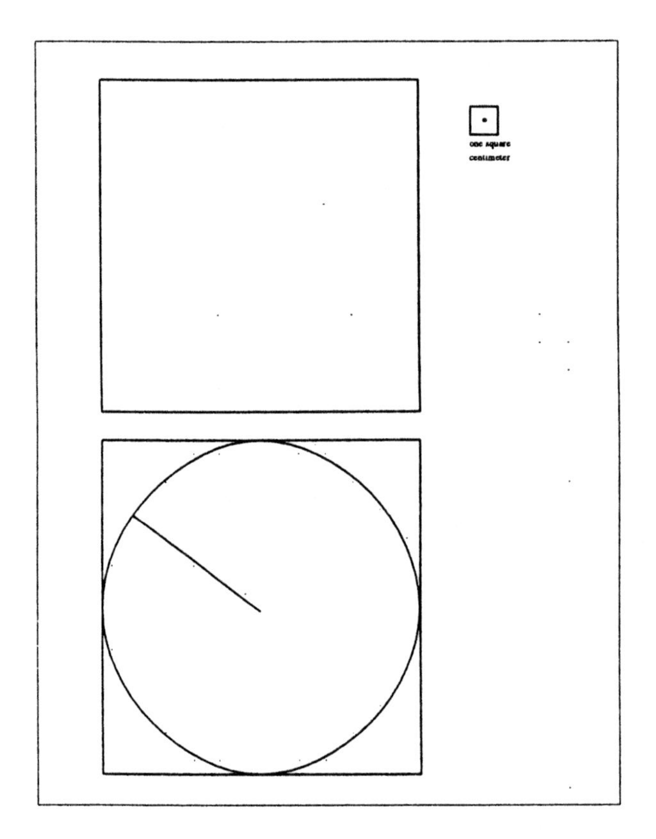

ILLUSTRATION 32. Handout for computing the area of a circle.

Row number:	Count:	Press:	Display:
9	10	[10][+]	90
10	10	[10][+]	100
11	8	[8][+]	108
12	4	[4][+]	112

The area of the circle is approximately 112 sq cm.

Second method: we can simplify the computations. From the first question we know that there are 144 dots in the whole square. In each corner of the square (outside the circle) we have eight dots. So inside the circle we have $144 - 4 \times 8 = 112$ dots ([4][*][8][+/−][+][144][=] or [144][M+][4][*][8][M−][MRC]).

Third method: the mathematical formula for a good approximation of the area of a circle with radius r is $3.14159 \times r^2$. (r^2 means "r squared" or "r times r.") Measure the radius (6 cm) and use the formula [6][*][=] [*][3.14159][=] (the display shows 113.09724). So the area is 113 sq cm. Our approximation was quite good.

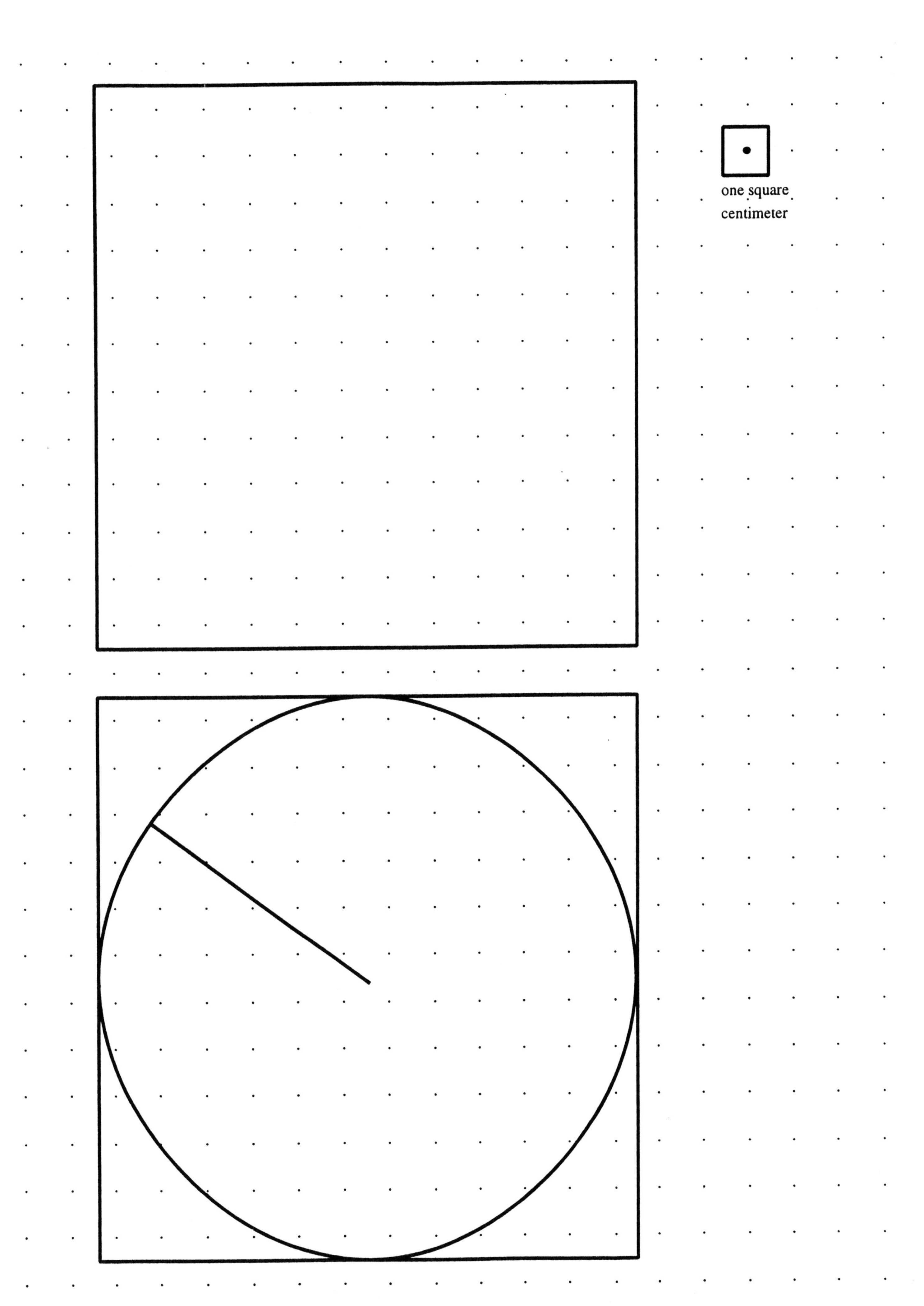
one square
centimeter

CHAPTER 42

Exploration of Stars

This lesson was taught in a fifth-grade class. The children had previously had some experience with calculators, but most had never used a compass, and many had never used a protractor. It is a lengthy lesson (about ninety minutes) and can be broken into two sessions if needed.

PROPS

- a blank sheet of paper per child
- six sheets of paper per child, each showing a large circle with a dot in the center (see handout at end of chapter)
- a ruler, protractor, compass, pencil, and calculator for each child
- transparencies of Pictures 1 through 6 (see Illustration 33) and six transparencies of a large circle with a dot in the center (like the children's) for the teacher
- a transparent ruler and protractor (for working on the overhead) and transparency markers
- a piece of string tied to a piece of chalk for drawing circles on the board, and a sheet of paper (like the children's) showing a circle with a dot in the center

THE LESSON

Before passing out the compasses (which have sharp points), I held one up and asked the children what it was. They knew that it was for drawing circles, but it took a while before anyone could say its name. I said, "We are going to use these today, but as you will see, they have sharp points. So before I hand them out will you pledge that you will use them responsibly—no poking or jabbing?" They responded yes. I also held up a protractor and asked what it was. Several knew its name. Both words, compass and protractor, I wrote on the board.

"Today's lesson is called 'Exploration of Stars.' Let me show you what we are going to make! Here are some pictures. What do you see in this one?" I showed Picture 1.

The kids didn't have to be asked twice. "That's a star."

"It's a nice even five-pointed star," said one. Another came to the overhead and drew a small (rather irregular) five-pointed star for the class, in one continuous stroke.

"How about this one?" I said, showing Picture 2, a six-pointed star.

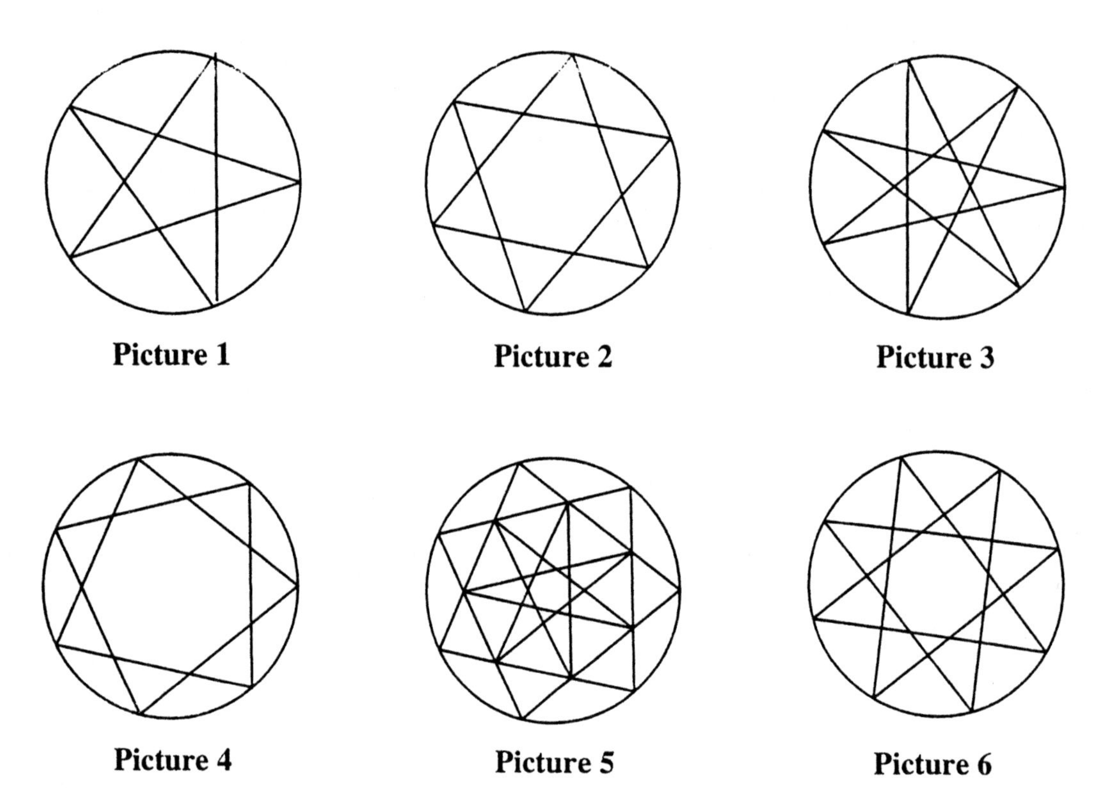

ILLUSTRATION 33. Stars with 5, 6, 7, and 8 points. Pictures 3 and 4 show two versions of a seven-pointed star. Picture 5 shows a skinny seven-pointed star inside a fat seven-pointed star.

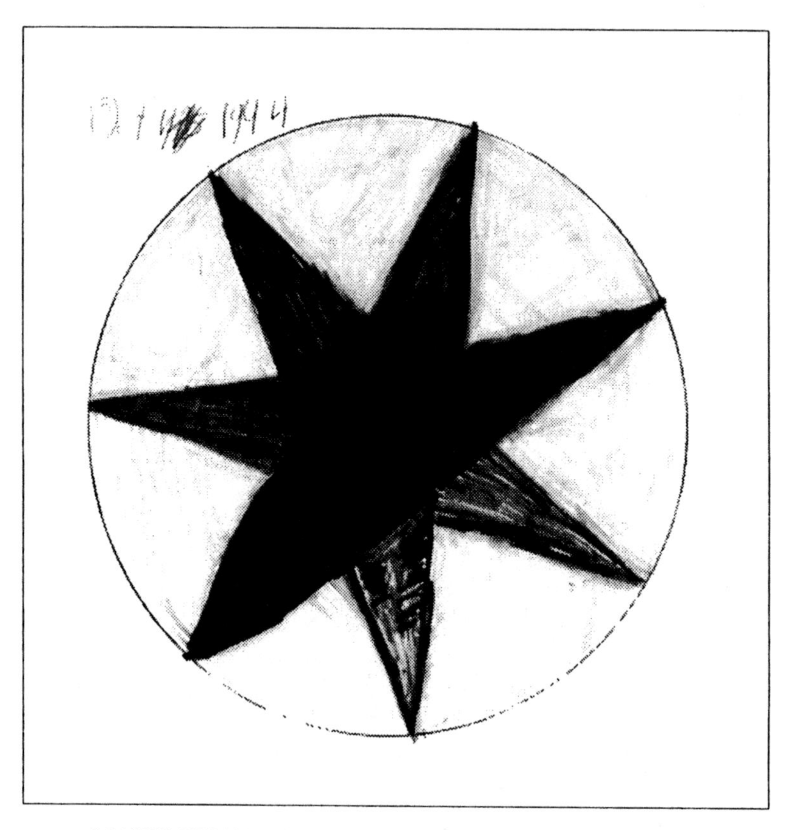

ILLUSTRATION 34. A fifth grader's drawing of a seven-pointed star.

"That's a Star of David!"
"There are two overlapping triangles."
"Are the triangles the same size and shape?" I asked.
"Yes!"
"Look at this one," I said, showing Picture 3.
"It has seven points!"
"And they are real skinny!"
"Here's another. What about it?" I asked, showing Picture 4.
"It has seven points, too!"
"But it is a fat star."
"Well, look what I can draw on the inside of the fat seven-pointed star," I said. I showed Picture 5.
"A skinny seven-pointed star!"
"Now look at this picture. What is it?" I asked, pointing to Picture 6.
"That star has eight points!"
"Would you like to make some of these?" I asked.
"Yes!"

Drawing Circles

"Before we start, let's take our blank sheet and our compass and just try to draw some circles. Can you do it?" Some children did not know how, but those who did soon began helping those who didn't. Pretty soon children had a page of circles. Some were concentric (having the same center); others overlapped. I showed the children how to draw a circle with a string and chalk. "What do you call a straight line that goes through the center of a circle and reaches to both sides?" (I drew one on my circle.) They did not quite know the word; one child said, "It is di-something!" I wrote "diameter" on the board and asked them to read it.

"Do you know the name for half a diameter, a straight line from the center of a circle to the edge?" The children did not know. I wrote "radius" on the board and asked them to read it.

Finding the Center of a Circle

"OK, before we begin to explore stars, I have a question. Do you see your six circles, each with a dot in the middle?"
"Yes," they replied.
"Well, how do you know that the dot is really in the middle? How can you prove that the dot is in the exact center?"
The kids had several ideas here. "You just draw a line across the circle and measure it."
"Well, how would that show that the dot is in the center?" I asked.
"The line needs to go through the dot."
"Is it enough that the line goes through the dot?" I asked.
"No! The dot could be near the edge of the circle."
"Well, suppose I gave you a circle that didn't have its center marked, and asked you to put a dot in its center. How would you do it?" I asked.
This was a hard question and brought consternation to the children's faces. While the children discussed ideas, I cut out my circle. "If we had more time, I'd ask you to do this,

too. But for now, I'll do it by myself. Just for fun, I'll cut out my circle and fold it in half.'' I folded and creased it. I then unfolded it and held it up.

''Oh,'' said several kids. ''That fold goes through the center.''

''Does it hit the dot?''

Several came to inspect and replied, ''Yes!''

''But we still don't know that the dot is really in the center.''

''Hey, wait a minute!'' said Cora. ''I think we can fold the circle again.''

''Into what?'' I asked.

''Into quarters!''

I asked Cora to come up and fold my circle and to crease it along the fold. ''Now unfold it and hold it up,'' I said.

''Is the dot exactly in the place where the two folds cross?'' asked Lance. Cora looked carefully and reported that, sure enough, it was.

The kids thought this was awesome – such a simple way to find the center of a circle. Just fold it into quarters and unfold it, and the center will be at the intersection of the two folds. Actually, another way to do it is to fold the circle in half, then unfold it and fold it again in half (along another diameter). Where the two folds meet is also the center of the circle!

Drawing Stars

''OK, are we ready to draw the first star?'' I asked. I put the five-pointed star transparency on the overhead.

''Yes!''

''Well, how are we going to do it?'' I asked.

''We just have to put dots on the circle's edge, and join the dots,'' said Ashley.

''We can use the protractor. See how you can put the hole in the protractor over the dot in the center of the circle,'' said Monica. I removed the star from the projector and put up a circle with a dot in the center. I placed the hole of the protractor over the dot.

''Uh oh,'' said several. ''the protractor does not reach to the edge of the circle.''

''Well, what do we need to do, anyway, to be able to draw the nice regular star?'' I asked.

Finally Jason piped up: ''We need to put five marks equally spaced on the edge of the circle.''

''Equally spaced?'' I asked. ''How are we going to do that?''

''We could use the protractor, but it is not big enough,'' said one.

''Well, if it were big enough, how would we be able to use it?'' I asked.

''We'd measure.''

''What would we measure?'' I asked. The children were not sure. I pointed to the scale on the protractor. ''Do you see these little lines on the edge? Do you know what the space between two lines is called?'' Silence. Mark said he knew how to draw it, but he didn't know what to call it. I wrote ''degrees'' on the board, and invited him to come up and make the symbol, °. ''How many degrees are shown on the protractor?'' I asked.

''170,'' said several.

''No, 180. You see how there are ten more, even though they are not labeled.''

''OK, and if I turn the protractor upside down, like this'' (I turned it), ''then how many degrees are there on this side?''

''180!''

''So how many degrees in the whole circle altogether?''

''360.''

''Can we use the degrees to help us divide up the circle?'' I asked. Jessica suggested we divide 360 by 5. "Let's use our calculators and see what we get." [360][÷][5][=] Display: 72.

"What does the 72 mean?" I asked.

"It means there are seventy-two degrees in one-fifth of a circle," said Jason.

"OK, let's draw a radius on our circle, but let's do it lightly, because later we want to erase it," I said. "Draw a straight line from the center to the edge. Everybody got it?"

"Just a minute!"

"OK," I said, "now line your protractor up with its hole over the center of the circle, and its 0° line lined up with the radius you just drew. Everybody got that?" I lined up my transparent protractor on the overhead. "Now, one of the points of the star will be at 0°. But where will the next one be?"

"Oh, I get it," said Stephen. "It will be at 72°."

"Cool," said one child.

"Let's make a mark at 72°, by the edge of our protractor."

"But," objected Cora, "we need a mark on the edge of the circle."

"Yeah, we do," I said. "Let's worry about that in a few minutes. Can everybody find 72° on your protractor? Put a mark there." I wrote 72° on the board.

"Now, where do you think the next mark goes?"

"You just add seventy-two more degrees," said Monica. "That's 144."

"OK, can we find 144° on out protractor and make a mark?" I asked.

"This is not so easy," said several. We needed to wait and help children until everyone had two marks on the edge of the protractor.

"How many points do we have now, around the edge of our protractor?" I asked.

"Two," said the children.

"Where shall we put the third mark?" I asked.

"It is 3 times 72," said Angela. "That is 216."

"216 what?" I asked.

"216 degrees," said Angela. "But I don't have 216 degrees on my protractor."

"There are several ways to do this," I said, "and you can use any method that is correct. I will show you one that is pretty easy." I turned my protractor over to the other side of the circle, and lined it up with my radius, putting the protractor's hole over the dot in the center of the circle. "Now I can start over from zero degrees, and I can make a mark at seventy-two degrees, as I did before on the other side. But this time I have to use the inner scale on the protractor. See these numbers inside? I want to make a mark at seventy-two degrees, but it must be measured using the inside numbers." I made a mark on my circle. "Does everybody get it? Who needs help?" There were lots of takers, and we had to spend a few minutes making sure that children got their protractors turned around and nicely lined up, and that they could read the inner scale.

"We are getting close now!" I said. "We need only one more mark, to have all our points marked. Where does it go?"

"It is just like on the other side," said Eddie. "The mark should be at 144°."

"Yes, but remember which scale to use. Is it the inside one or the outside one?" I asked.

"Inside," said Eddie.

I offered help to those having difficulty, and after a few minutes children had their four marks. "Before you move your protractor," I said, "just make a little mark at 0°. That mark will be on the radius that you drew. Can you do it?" I made a mark on the overhead and went around to check that children had five points marked on the inside of their circles.

"Now look at your circles. Can you imagine a five-pointed star inside?" Several said they could.

"But what we need are marks on the edge of the circle," said several. "And what we have are marks on the inside."

"OK, that is the challenge. How are we going to get marks on the edge of the circle?" I asked.

There was a long pause, as children looked at their marks. Finally Cora said, "Well, we have one mark on the edge. We made it with a line from the center to the edge, through a mark at zero degrees. Maybe we could draw some more lines from the center to the edge."

I asked Cora if she would come to the overhead and show us what she meant. She lined up a ruler with the center and a mark in the inside of the circle and made a mark on the edge of the circle. She continued to do this until there were five marks on the edge.

"Oh, I get it," said several.

"Try it," I said. "If anyone needs help, let me know." The children went to work, but it was not easy. A few could do it quickly, and I asked them to help others. Finally, everyone had five marks on the edge of his or her circle.

"Are we ready to draw our stars now?" I asked.

"Yes!"

I put Picture 1 back on the overhead. "Try it," I said. "Do you know which marks to connect?"

"Yes," said Gerald. "You skip a mark. You connect every other mark."

Children needed help, but eventually everyone had a five-pointed star. They agreed that it was a lot of work, but it was worth it. I suggested that they erase the radius they had drawn, since it was now longer needed.

"Do you want to try a six-pointed star?" I asked.

"Yes," they said.

"OK, let's take a new sheet and first lightly draw a radius. Now let's line up our protractor. Where do we put our first mark?"

"It's 360 divided by 6," said several. [360][÷][6][=] Display: 60.

"The first mark goes at 60°," said one. I wrote 60° on the board.

"Where do the next points go?" I asked.

"Easy," said several. "At 120° and 180°." Children then flipped their protractors over, and marked 60° and 120° on the other side. They did this very quickly. They then made marks on the circle's edge by lining a ruler up with the center of the circle and the marks inside the circle. Before I knew it, six-pointed stars began to appear. I was amazed at their speed and proficiency, and by their obvious delight.

"Look, I drew the two triangles!" said one.

"Yes, but can you see all the other triangles?" asked another. Yet another child showed me a star within a star that she drew.

"Does anyone want to draw a seven-pointed star?" I asked. All but one child said yes. But it was time to go to recess.

"What do you want to do?" asked the teacher. One child proposed that they stay inside now and work on stars and take a double recess in the afternoon. This was voted on and approved by the children. So we began the seven-pointed stars.

"First we need to draw a radius lightly and get our seven marks. But where does the first one go?" I asked. [360][÷][7][=] Display: 51.428571.

The children had no difficulty interpreting this. They said the first mark should go at 51° or 52°. I wrote it on the board. "Let's figure out where the second mark should go," I said.

"Easy," said several. "It's 102° or 103°."

"OK," I said. I wrote that on the board also. "What about the third mark? I'll show you the calculator button presses if you want.'' [360][÷][7][*][3][=] Display: 154.28571. "What should we mark?" I asked.

"154°," answered a few children. I wrote it on the board.

"OK, when you get those marks, let's flip the protractor over and try again, remembering that we need to use the inner scale. Where do the marks go?" Jason said they should be put according to the list on the board: at 51°, 102°, and 154°. "When you get the seven marks, then make your seven marks on the circle's edge," I said.

I checked to see that the children were following and noted that some were already ahead. I showed on the overhead the two seven-point stars, the skinny one and the fat one (Pictures 3 and 4). "Which star do you want to make?" I asked. "Do you see how they are different?" Ashley said she wanted to make a skinny star, and she knew she had to skip two marks on the circle's edge. Lance made a fat star, and he knew he had to skip only one mark.

It was time for me to go, so I left the class, which was still a beehive of activity. Children were making stars and beginning to color them. They made extra stars in the middle (as in Picture 5) also. They told me they liked the lesson and asked me, "Is this math?"

Before I left the school that day, I went by the principal's office. She said, "Please go back to the class where you taught the star lesson. The children have something for you!" I returned, and the children had (finally) gone out to recess. But the teacher presented me with a beautifully colored seven-pointed star, drawn by Cora, for me to take home (see Illustration 34). She said it was the only one she could get to present to me; the children wanted to take the others home to show their families. It was a wonderful moment for me.

REMARKS

(1) Which star should the lesson begin with? Would it have been better to start with the six-pointed star, since it was easier to make? Maybe not. After the children had made five-pointed stars with a lot of help, many of them were able to make a six-pointed star by themselves. This would not be the case if we had started with the easier star first. The approach I used was to show or help the children through a fairly complex generic case, which gives a correct general view of the problem, and then to see if they can do an easier case for themselves.

(2) If time had permitted, I would have introduced the words "circumference" (instead of edge) of a circle, and "chord," for a straight line drawn across a circle, not necessarily going through the center.

(3) The lesson clearly required more time than we had.

(4) This lesson might be followed by another, in which children investigate mathematical properties of stars, including:
   - When do stars split into two parts, similar to the way that the six-pointed star splits into two triangles? (It depends on the divisors of the number of points. Here $6 = 2 \times 3$. For every number greater than six, there is a star that does not split.)
   - Measure the angles at the points. What part of the whole angle (360 degrees) do they form?
   - Measure the chords and take the ratio of a chord to the diameter of the circle. For which star is this ratio the smallest? For which is it the largest?

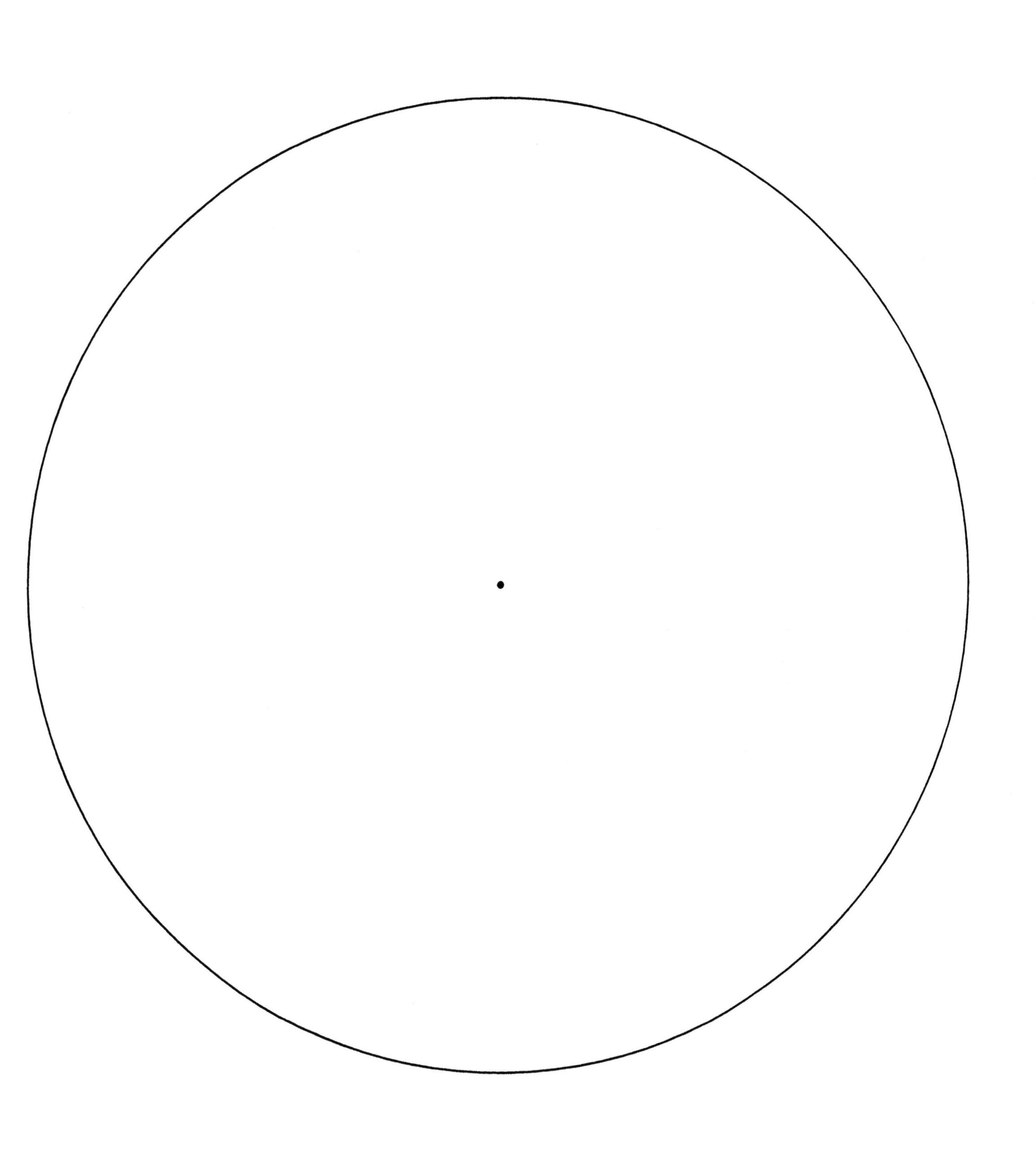

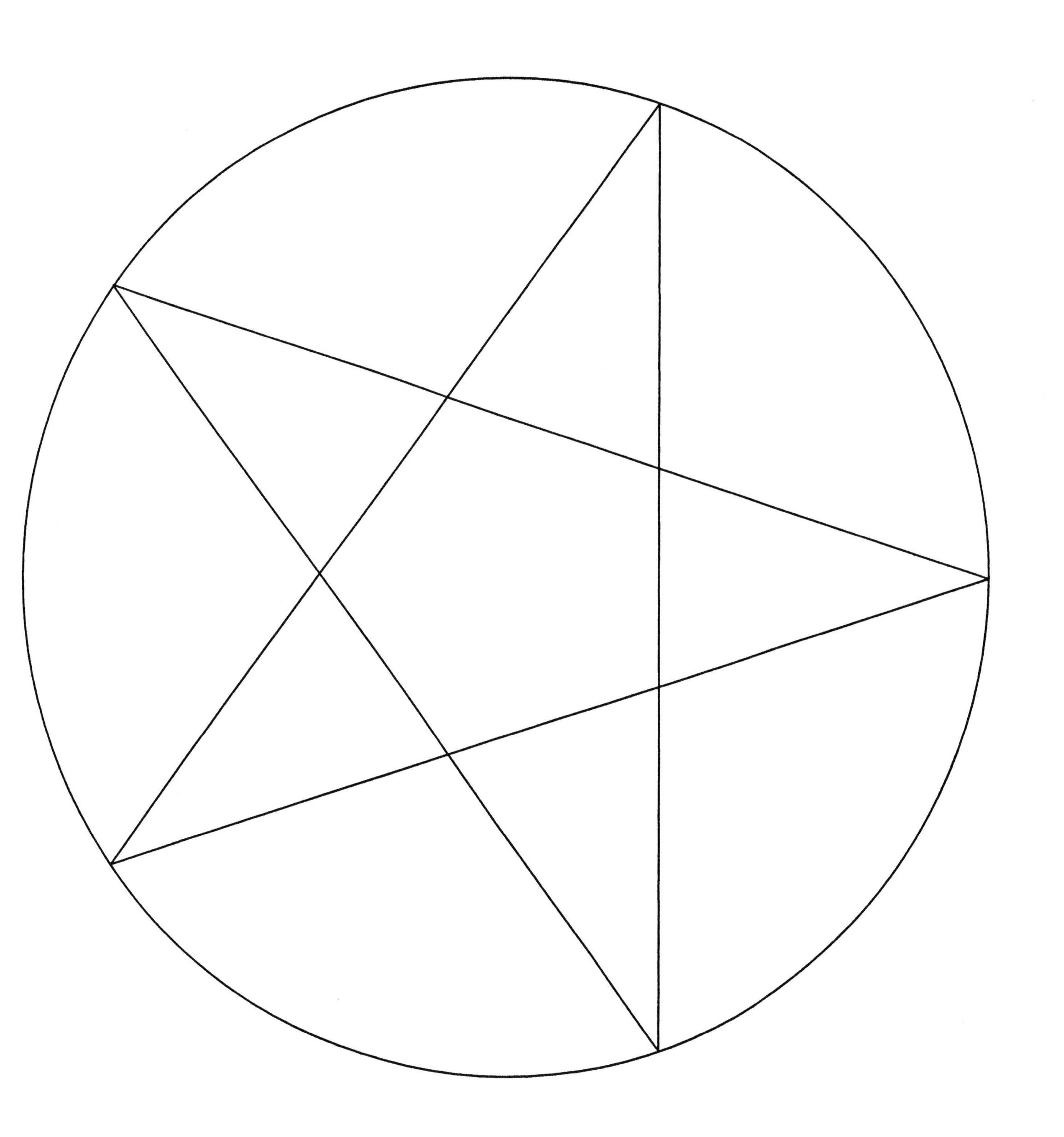

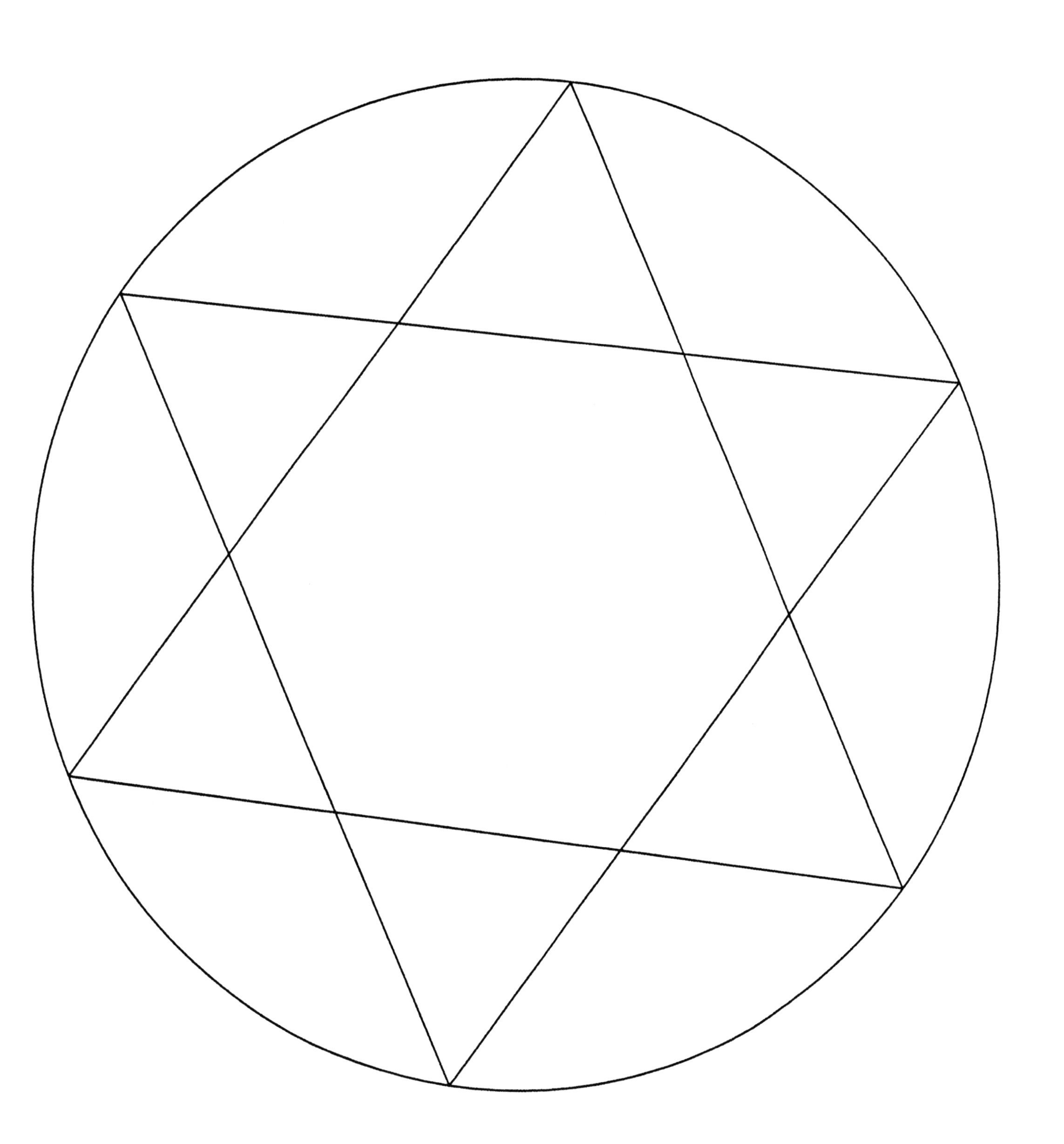

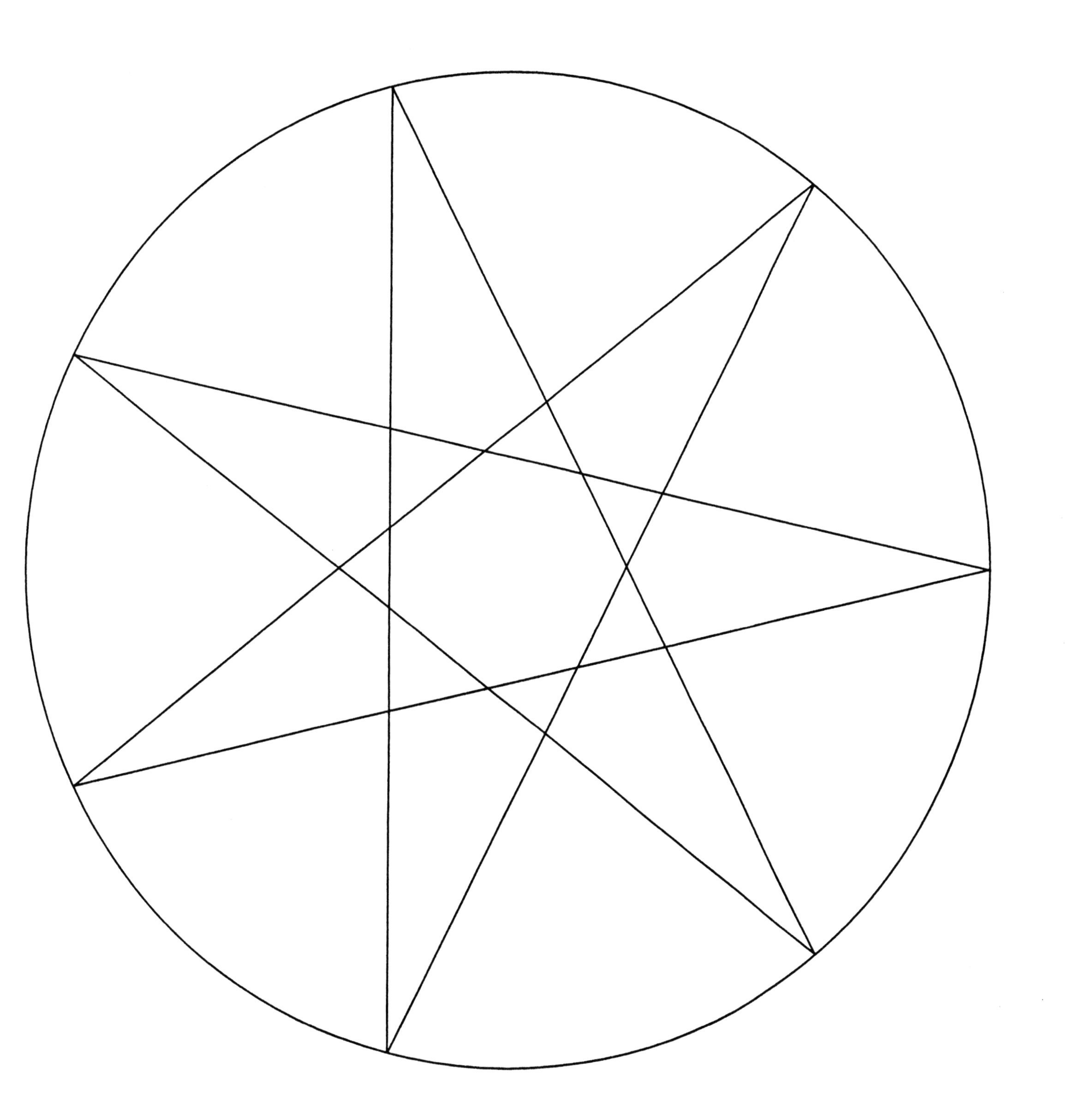

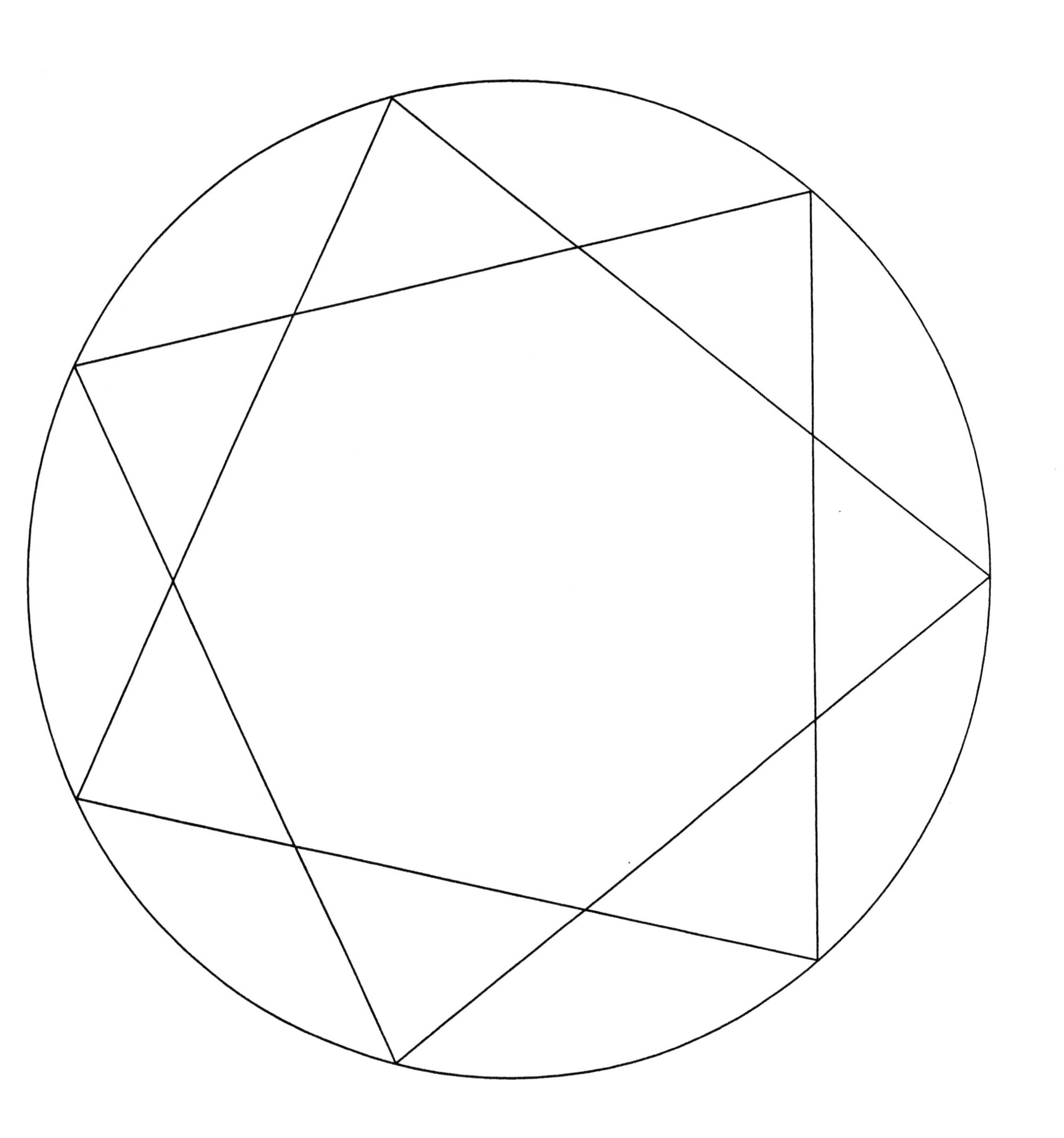

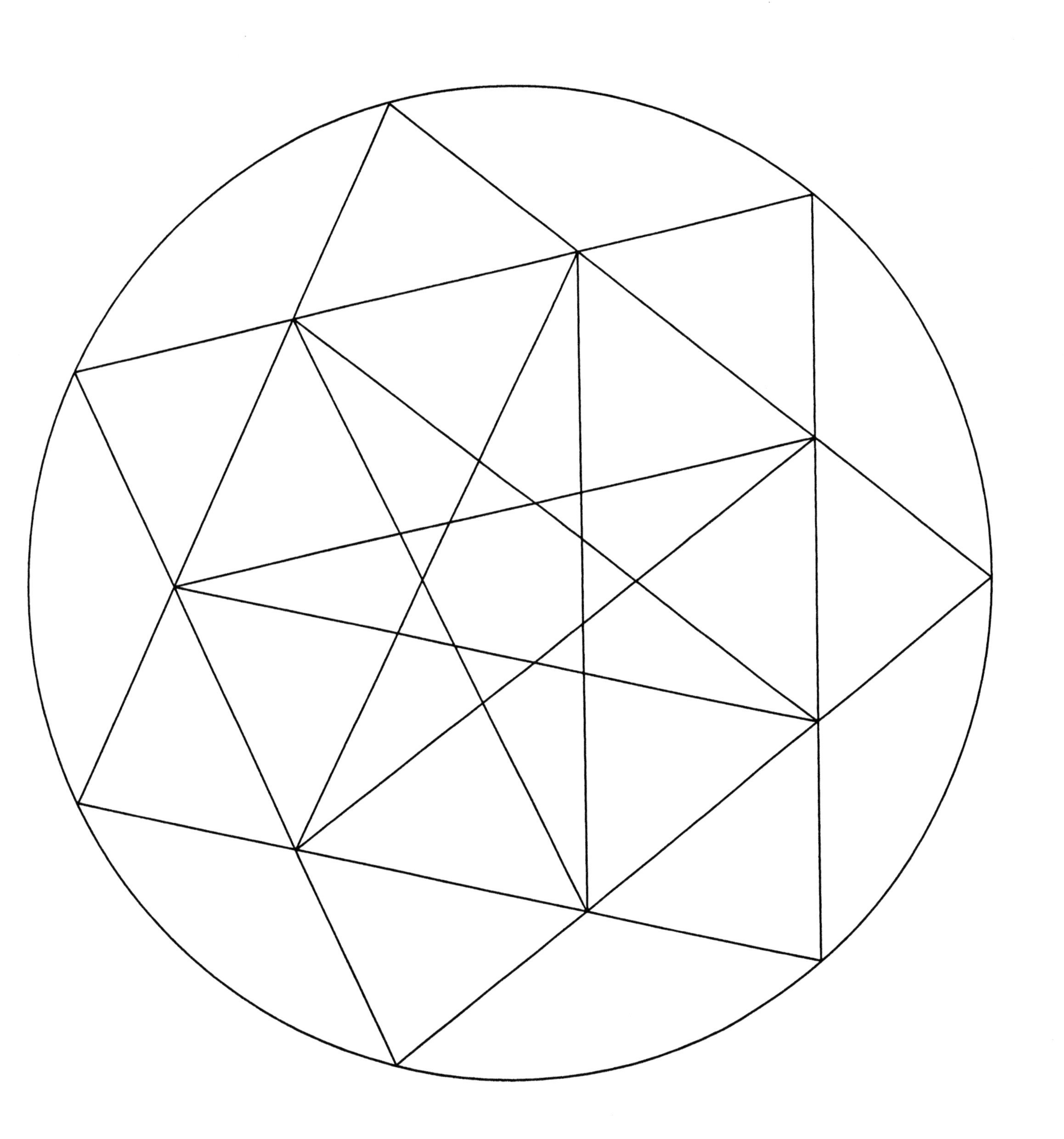

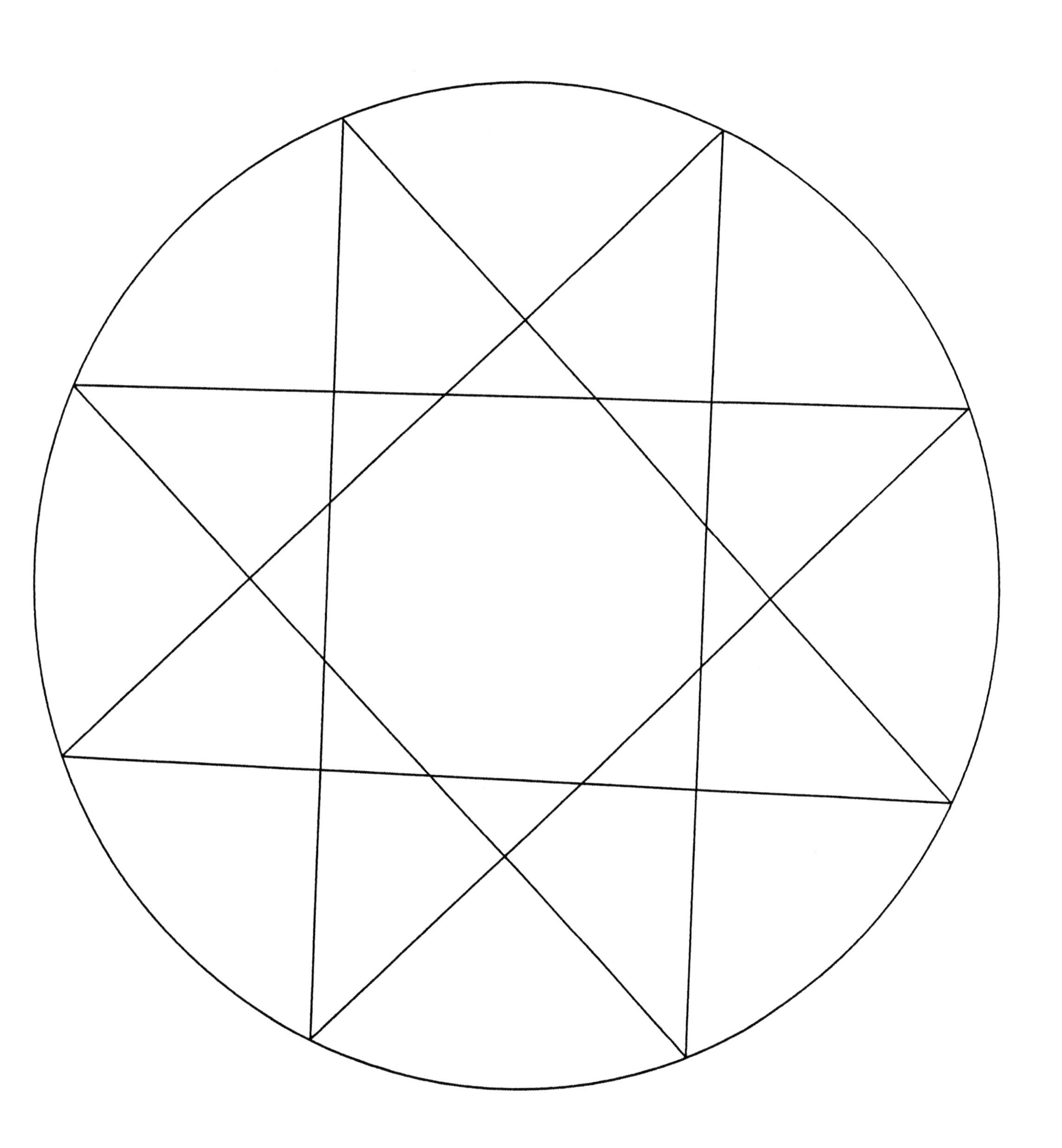

CHAPTER 43

Eclipse of the Sun (Recommended for Sixth Grade and Higher)

An eclipse of the sun occurs when the moon gets between the earth and the sun. The shadow of the moon on the earth makes the eclipse (see Illustration 35). If the sun is completely covered, the eclipse is total, but from most places on earth a part of the sun is still visible, and the eclipse in these places is called partial. The disks of the moon and the sun on the sky appear to be the same size. In reality the disk of the sun changes its size slightly because the distance from the earth to the sun varies during the year. (This is not the reason for seasonal changes!) In our calculations we will assume that the sun's and the moon's disks are equal in diameter (see Illustration 36). Illustration 37 shows six stages of a partial eclipse. In the fourth stage of the picture, the sun is almost covered. But even then, it is not dark on earth, because human eyes adjust very well to the smaller amount of light.

PROPS AND TOOLS

- for each group of two to four, rulers, protractors, calculators, and a picture of one stage of an eclipse (Three phases are included with this book; see end of lesson.)

THE LESSON

The task is to compute the percentage of the sun's disk that is covered by the moon.

Example (see Illustration 38)

(1) Drawing: find the center of the sun's disk C. It can lie inside or outside the darker area. Bisect the shaded area by the chord AB. Draw triangle ABC and its height.

(2) Measuring: measure the radius of the circle, the chord AB, and the height of triangle ABC in centimeters and millimeters. Write down the measurements. Measure the angle ACB in degrees.

(3) Planning computations: half of the shaded area is the difference between section CAB of the circle and the triangle CAB. The area of the section = (angle/360) × pi × radius^2. The area of the triangle = chord × height/2. The area of the sun's disk = pi × radius^2. So we have to compute: 2 × (area of the section − area of the triangle)/area of the sun's disk and express the result as a percentage.

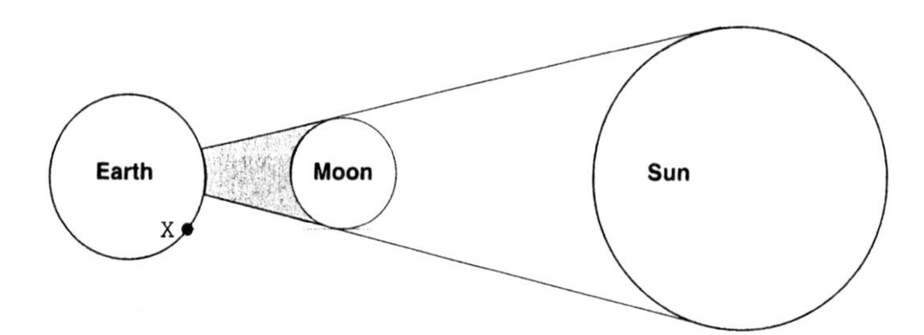

ILLUSTRATION 35. A person standing at point X sees the sun partially covered (not to scale).

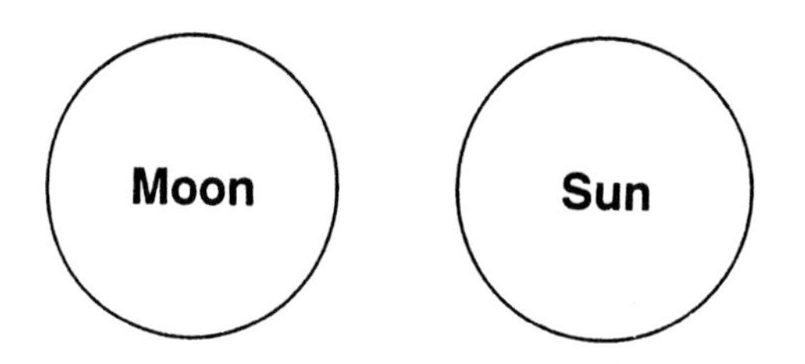

ILLUSTRATION 36. From the earth, the sun and the moon appear to be the same size.

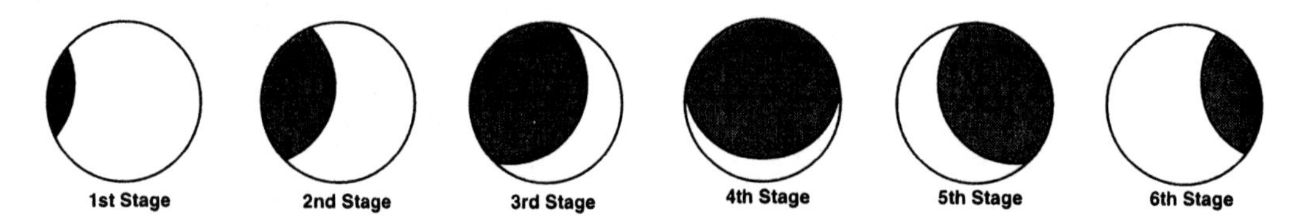

ILLUSTRATION 37. Six stages of a partial eclipse.

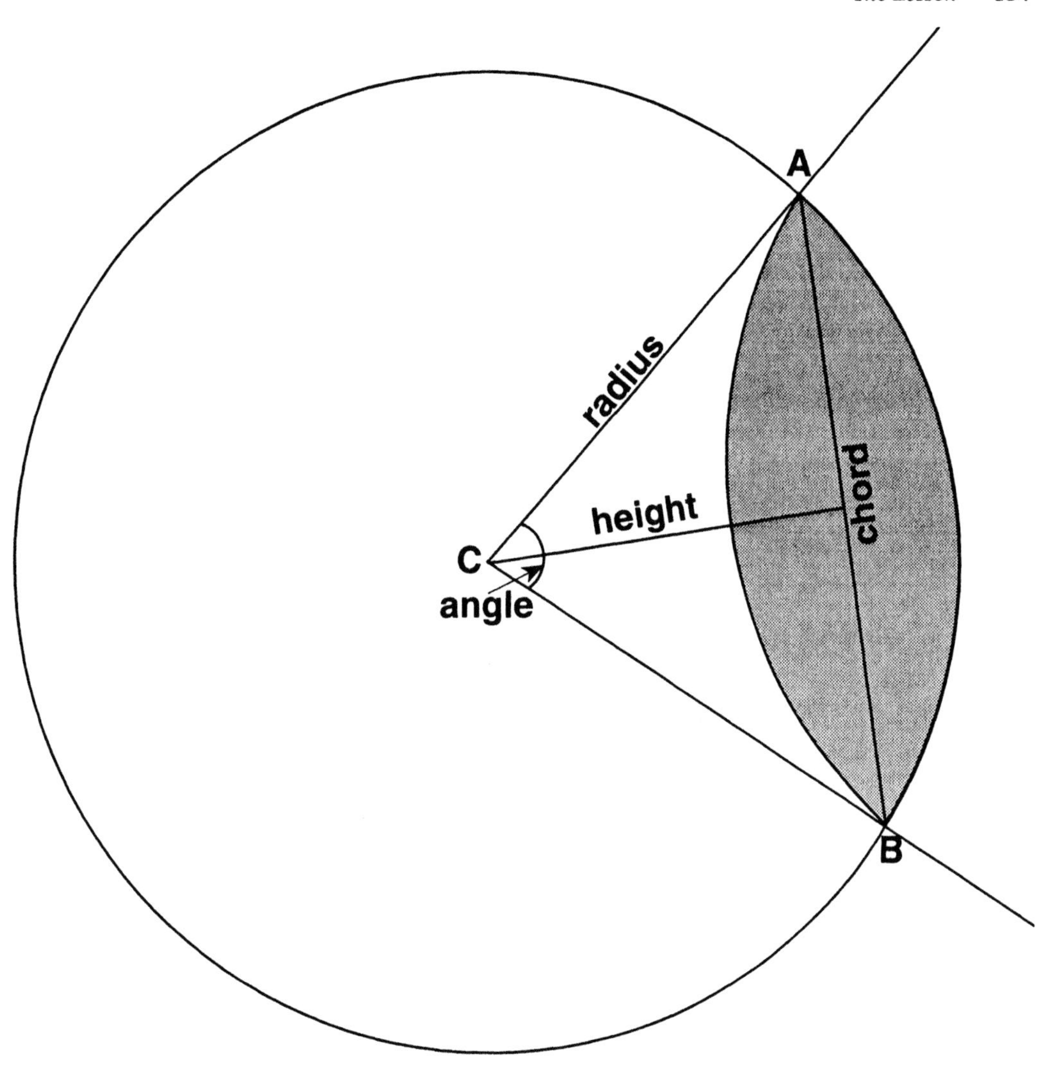

ILLUSTRATION 38. A drawing used for computing the percentage of the sun covered by the moon.

(4) Writing a program: The formula above can be simplified to:

$2 \times (\text{angle}/360 - \text{chord} \times \text{height}/(2 \times \text{pi} \times \text{radius}^2)) = \text{angle}/180 - \text{chord} \times \text{height}/3.14/\text{radius}/\text{radius}$.

Generic program:

Keystrokes:	Comments:
[chord][*][height][÷][3.14][÷][radius][=][=][M+]	Divide by the radius twice.
[angle][÷][180][−][MRC][=]	Compute the formula.
[÷][1][%]	Change it into a percentage.

(5) Computing:
Data (based on Illustration 38):
chord = 8.4 cm
height = 5.0 cm
radius = 6.5 cm
angle = 81°

Keystrokes:	Display:
[8.4][*][5][÷][3.14][÷][6.5][=][=][M+]	0.3165868
[81][÷][180][−][MRC][=]	0.1334132
[÷][1][%]	13.34132

Answer: 13% of the sun's disk is covered.

REMARKS

The method and results should be thoroughly discussed. Examples of questions: How did you find the center of the disk? How did you bisect the shaded area? How did you derive the formulas? Can you show us that they are correct? How did you design the program?

CHAPTER 44

Square Pizza

Children should be familiar with measuring perimeters of polygons and comparing areas of figures having different shapes.

The formula for the area of a triangle: base × height/2.

TOOLS AND PROPS

- rulers
- scissors
- square pieces of paper with colored boundary (for make-believe pizzas with special crust–we used origami paper)

THE LESSON

A new pizza place has opened in [your town's name]. It bakes very good pizza with excellent crust. But it bakes only square pizzas–no round ones!

One day Andy, Barb, and Chuck got a pizza for lunch. They wanted to cut it into three equal portions, but also each one wanted to have the same amount of crust. Can you help them? (For the purposes of this problem, let's say that the crust is simply the boundary of the square.)

Andy said that they cannot all have pieces which have the same shape, but they all agree that it is okay as long as each person has a fair share of the whole pizza and a fair share of the crust.

Barb said that it would be easy to cut the pizza into four parts of the same shape and size, and having the same amount of crust.

Chuck said that dividing a square pizza into four parts is easy; he can do it in two different ways. Barb said that it is not a big deal, and that she can do it four different ways (see Illustration 39).

Then Andy said that he knows how to divide the pizza into three parts (Illustration 40).

Mathematical Formulation of the Task

Divide a square into three parts, such that they all have equal areas and contain equal lengths of the square's boundary.

ILLUSTRATION 39. Square pizzas divided into four parts of the same shape and size, and having the same amount of crust.

Suggested Method

Guided discovery.

Mathematical Principle Involved

If we cut a section of a square from the center, then the area of the section is proportional to the length of the boundary of the square (crust) lying on this section. So to make fair division into any number of pieces, it is enough to divide the boundary into parts of equal length, and then cut from the center. This method works for any polygon in which one can inscribe a circle (a polygon in which all sides are the same distance from the center). This is a simple application of the formula for the area of a triangle. A section can be viewed as consisting of one or more triangles with the same heights and bases lying on the boundary of a polygon.

REMARKS

In spite of its simplicity, this reasoning requires a considerable amount of mathematical sophistication, and it should not be attempted without the prerequisites specified above.

ILLUSTRATION 40. Square pizzas divided into three equal parts (not of the same shape, but having the same amount of crust.

CHAPTER 45

What is Fair?

In the ancient city of Basra, two friends, Mustapha and Ali, sat down in the shade to eat their lunch. Mustapha brought five slices of bread, but Ali brought only three slices.

Suddenly, a stranger approached and said, "I am new in this town and I am hungry. Will you share your lunch with me?"

Mustapha and Ali shared their lunch. They broke some slices of bread into smaller pieces so everyone ate exactly one-third of eight slices.

After thanking them, the stranger said, "To repay you for your hospitality I am giving you eight silver pennies."

Then Ali said to Mustapha, "I brought three slices of bread, and you brought five. So I will have three pennies and you will have five pennies."

"Wait a moment!" Mustapha answered. "Let's see how much bread each of us gave the stranger. You brought three slices, but you ate eight-thirds which is two and two-thirds slices. So you gave the stranger only one-third of one slice! On the other hand, I gave the stranger five slices minus the two and two-thirds slices I ate, which is seven one-third slices. So I should get seven pennies and you only one!"

Then the stranger said, "The money is yours, and you may divide it any way you want, but I have a suggestion. Because both of you shared your lunch with me, each of you deserves four pennies, regardless of how much bread you brought, and how much of it I ate."

Question: How should they divide the eight pennies?

Comment: Arithmetic answers the questions "How much?" and "How many?", but it does not answer the question "What is fair?"

CHAPTER 46

Getting Familiar with the Calculator Keys

Different brands of calculators have slightly different key arrangements, and some keys can have slightly different functions. This description is based on the TI-108 calculator.

The [ON/C] (on/clear) key has several functions:

(1) Press it to turn the calculator on.
(2) Press it to restart the calculator after an error message appears (*E* in the left-most column of the display).
(3) Press it once to clear the last entry.
Example: I wanted to add 4.56, 3.24, and 5.75. I pressed
[4.56][+][3.24][+][575].
What should I do? Press [ON/C][5.75][=].
(4) Press it twice to clear everything except the calculator's memory.

There are other two-key combinations that are important:

(1) Pressing [MRC] (the memory recall) key twice clears the calculator's memory.
(2) Pressing [*][=] (the times-key followed by the equal-key) computes the square of the displayed number.
(3) Pressing [÷][=] (the division-key followed by the equal-key) computes the reciprocal. Example. Press [5][÷][=]. You see 0.2 (which is 1/5, the reciprocal of 5). Press [÷][=] again. You see 5.

Remember that when you perform a chain of operations such as [1][+][2][*][3][−][4][=], they are executed from left to right, sequentially, without precedence. So you compute $(1 + 2) \times 3 - 4$ (answer: 5), and not $1 + (2 \times 3) - 4$ (answer: 3).

At a given time, there are three numbers in a calculator. One is in the memory, a second one is on the display, and a third one is hidden. If you press [2][+][5], then 5 is on the display and 2 becomes hidden. If you now press [=], the result, 7, appears on the display, and 5, the second argument, becomes hidden. If you press [=] again, and again, you get 12, 17, 22, . . . and 5 will remain hidden. This also works for subtraction, division, and multiplication, with one exception for multiplication. If you press [2][*][5][=] you see 10, but the *first* argument, 2, not the second, becomes hidden. So if you press [=] again, you see 20 $(2 \times 5 \times 2)$, not 50 $(2 \times 5 \times 5)$.

If you want to enter a negative number such as −15, press [15][+/−] (15 followed by

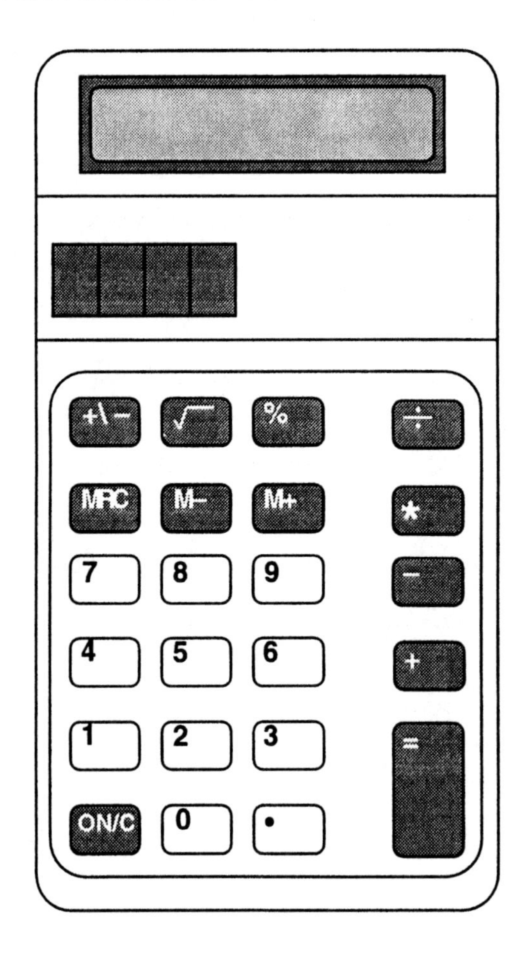

ILLUSTRATION 41. Getting familiar with the calculator keys.

the change of sign key, also called the plus minus key). If you press [+/−] again, you get 15 back.

Pressing [M+] (memory plus) and [M−] (memory minus), adds and subtracts the displayed value to the memory.

The [√] key (square root) computes the square root of the displayed number. If you try to compute the square root of a negative number, you get an error message.

When computing percents, you do not use the equals key at all. To compute 10 percent of 12.75 you press [12.75][*][10][%]. To add 10 percent of 12.75 to 12.75, you press [12.75][+][10][%].

PRACTICE EXERCISES (WITH ANSWERS)

Using calculators requires a considerable amount of skill. Two common problems are pressing the wrong key and pressing a key so lightly that the result does not register on the display.

Other problems are forgetting the decimal point, forgetting to clear the memory, and misreading multi-digit numbers from the display. In the early stages of learning, the best approach is not to hurry and to look at the display often, not just at the end when you are getting the final result. It takes some concentration. But paying too much attention to the process of calculation carries a price. Computation is not a goal in itself, and it occurs within a general framework of problem solving. Concentrating on the process of computation detracts attention from other aspects of the problem at hand. The ultimate goal is to be able to perform all computations quickly, correctly, and with a minimal amount of attention. Such a goal can be achieved only after several years of constantly using a calculator, if ever. From the very beginning, attention should be paid to manual skills, because high error rates make more complex uses of calculators practically worthless.

Below is a set of exercises. If you can do them correctly without too much effort, fine. Otherwise some additional practice is recommended.

(1) Add the lists of numbers. (Do not copy the lists onto paper; just punch in the numbers.) Add each list twice, once top down and once bottom up, before you look up the answer, which is given at the end.

123.25	0.0125	129076.01
86.12	0.981	231.25
570.49	0.3541	0.1231038
89.50	2.0009	78432.2
34.98	1.12	28.33333
723.26	0.8046	0.0009743

(2) Add the lists of numbers. (Do not copy the lists.)
(a) 31, 14, 16, 92, 26, 53, 58, 19 sum = ________________
(b) .74, 1.5, 3.2, 7.1, 10, 1.8, 3 sum = ________________

(3) Add and subtract the fractions. For example,
1/5 − 2/3 + 6/7 = ____________________
Keystrokes: [5][÷][M+] [2][÷][3][M−] [6][÷][7][M+] [MRC]
Answer: 0.3904762
Remark: [5][÷][M+] has the same effect as [1][÷][5][M+].
(a) 2/3 + 3/4 + 4/5 + 5/6 − 7/8 − 8/9 = ________________
(b) 1/2 − 1/3 + 1/4 − 1/5 + 1/6 − 1/7 + 1/8 = ________________

(4) Read each problem aloud, compute its answer, and read aloud the answer on the display. (Do not write down the answers.)
(a) 765.23 × 46.4
(b) 1228 × 9876
(c) 0.73562 ÷ 0.27

Answers

(1) (a) 1627.6, (b) 5.2731, (c) 207767.91

(2) (a) 309, (b) 27.34

(3) (a) 1.2861111, (b) 0.3654762

(4) (a) Seven hundred sixty-five and twenty-three hundredths times forty-six and four tenths equals thirty-five thousand five hundred six and six hundred seventy-two thousandths.
(b) Twelve hundred twenty-eight times nine thousand eight hundred seventy-six equals twelve million one hundred twenty-seven thousand seven hundred twenty-eight.
(c) Zero point seven three five six two divided by zero point twenty-seven equals two point seven two four five one eight five.

GUESS AND CHECK (WITH ANSWERS)

If you want to improve your calculator skills, you may try these problems. To solve them you may try to imagine what is happening inside the calculator. These exercises are prepared for the TI-108 calculator and may not work for other calculators.

Guess the answer which will be shown on the calculator's display, then press the keys to see if you are right. Clear the calculator before starting the next problem.

Keystrokes:	Display:
(1) [2][+][3][*][4][=]	____________
(2) [3][+/−][+][3][=]	____________
(3) [5][M+][M+][MRC]	____________

Remember to clear the calculator!

(4) [5][M+][[M−][1][÷][MRC][=]	____________
(5) [10][*][=]	____________
(6) [10][*][=][=]	____________
(7) [100][÷][=]	____________
(8) [.01][+/−][÷][=]	____________
(9) [100][√]	____________
(10) [2][√][*][=]	____________
(11) [200][−][15][%]	____________
(12) [100][−][10][=]	____________
[=]	____________
[=]	____________
(13) [100][+][3][=]	____________
[200][=]	____________
[1][=]	____________
(14) [10][−][3][M+]	____________
[MRC]	____________
(15) [1][+][100][√][=]	____________
(16) [1][M+][*][MRC][÷][MRC][−][MRC][=]	____________

Answers

(1) 20. Operations are performed in order, without any precedence. $(2 + 3) \times 4$ was computed.

(2) $(-3) + 3 = 0$.

(3) 10. Memory was empty. This means that it contained 0. So after [5][M+][M+], $0 + 5 + 5 = 10$ is in the memory, and this value is displayed.

(4) Error: after [5][M+][M−], memory contains $0 + 5 - 5 = 0$. So [1][÷][MRC][=] attempts to divide 1 by 0, which creates an error.

(5) 100. The combination [*][=] (also written as [*=]) computes the square of the displayed number. Reading this sequence of keystrokes as "ten square" is preferable to "ten times equal."

(6) 1000. [*][=][=] computes the cube of the displayed number.

(7) 0.01. [÷][=] computes the reciprocal of the displayed number (here, 1/100). The reciprocal can also be called inverse. Reading "100 inverse" is preferred to "100 divided equal" and even more preferred than "100 divided by equal," which is nonsensical and therefore misleading.
(8) −100. The reciprocal of −0.01.
(9) 10. The square root of 100.
(10) 1.9999998. It should be 2, because you computed the square of the square root of 2. But the value of the square root of 2 was rounded down, which created a small error.
(11) 170. 200 minus 15% of 200 equals 200 − 30 equals 170.
(12) 90, 80, 70. We are counting down from 100 by 10. 10 stays in the hidden register, and the calculator also remembers that the operation is −.
(13) 103, 203, 4. Now 3 stays in the hidden register and the calculator remembers that the operation is +.
(14) 7, 7. The calculator first performs the subtraction and then adds the difference to the memory.
(15) 11. The calculator first computes the square root of 100, and then adds it to 1.
(16) 0. It is simply 1 × 1 ÷ 1 − 1, which is 0.

PROBLEMS (WITH ANSWERS)

Solve the following problems using a calculator. Write down the sequence of keystrokes that you use and compare it to the answers given on the answer page. Usually a problem can be solved in many different ways, so if your answer is correct but your keystrokes are different, it means that you used a different method. Clear the calculator before starting the next problem.

(1) Add the numbers: 5.3, 7.6, 8.5.
Keystrokes:

____

Answer: __________________________________

(2) Add the numbers: −3.23, 123, −1.78.
Keystrokes:

____

Answer: __________________________________

(3) Compute the value of 7 × 2.99 + 8 × 7.49.
Keystrokes:

____

Answer: __________________________________

(4) Count down from 20 to −20 by 5s.
Keystrokes:

____

(5) Show the powers of 2 (4, 8, 16, . . .) up to 1024.
Keystrokes:

____

(6) You want to compute 123.45 ÷ 7.56. You have already pressed [1][2][3][.][4][5][÷][7][5] and you notice that you forgot to press the decimal point key. What should you do?
Keystrokes:

____

Answer: ______________________________

(7) Compute 37^2 (37 square).
Keystrokes:

____

Answer: ______________________________

(8) Compute 1/987 (the reciprocal of 987).
Keystrokes:

____

Answer: ______________________________

(9) Compute 25% of 175.
Keystrokes:

____

Answer: ______________________________

(10) Decrease 75 by 10%.
Keystrokes:

____

Answer: ______________________________

(11) Decrease 60 by 20% and increase the result by 30%.
Keystrokes:

____

Answer: ______________________________

(12) Compute the reciprocal of the square root of $7.7^2 + 8.8^2$ and store the result in the calculator's memory.
Keystrokes:

____

____

Answer: ______________________________

Answers

(1) [5.3][+][7.6][+][8.5][=] Answer: 21.4
(2) [3.23][+/−][+][123][+][1.78][+/−][=] Answer: 117.99
(3) [7][*][2.99][M+][8][*][7.49][M+][MRC] Answer: 80.85
(4) [20][−][5][=][=][=][=][=][=][=][=]
(5) [2][*][=][=][=][=][=][=][=][=][=]
(6) [ON/C][7][.][5][6][=] Answer: 16.329365
(7) [37][*][=] Answer: 1369
(8) [987][÷][=] Answer: 0.0010131

(9) [175][*][25][%] Answer: 43.75
(10) [75][−][10][%] Answer: 67.5
(11) [60][−][20][%][+][30][%] Answer: 62.4
(12) [7.7][*][M+] [8.8][*][M+]
[MRC][MRC][√] [÷][=] [M+] Answer: 0.08552

CORRECTING KEYSTROKE ERRORS ON THE CALCULATOR

When you press a wrong key on the calculator, you still can correct the error most of the time. It is important to know how to correct an error when you are performing long computations, because starting from the beginning again is annoying and time consuming.

(1) If you enter a wrong number, press [ON/C] once (you will see 0. on the display) and enter the correct number.
Example:
I want to compute (12 + 17) ÷ 34 + 5.

Keystrokes:	Display:	Comments:
[12][+][17][÷][43]	43.	I noticed an error.
[ON/C]	0.	I removed 43.
[34]	34.	The error is corrected.
[+][5][=]	5.8529411	

If you press an operation key, the memory key, or the equals key, then it is too late to correct the entry. So look at the display to be sure it is correct before pressing these keys.
Pressing [ON/C] twice clears everything except memory. You may correct an error more than once.

Example:

Keystrokes:	Display:
[10][−][123]	123
[ON/C]	0
[.123]	0.123
[ON/C]	0
[1.23]	1.23
[=]	8.77

(2) If you enter the wrong operation, simply press the correct one immediately after (do not press [ON/C]).
Example:
I want to compute (15 + 18) × 7.

Keystrokes:	Display:	Comments:
[15][+][18][+]	33	I made an error.
[*]	33	The error is corrected.
[7][=]	231	

Because you do not see the operations on the display, look at the keyboard when pressing operation keys.
Again you may correct an error many times.

Keystrokes:	Display:
[10][+][−][*][5][=]	50

(3) If you forgot to clear the memory, and you already have a number on the display that you do not want to lose, press [*][MRC][MRC][1][=].
Example:

Keystrokes:	Display:	Comments:
[77][M+]	M 77	77 is also in memory.
[1994]	M 1994	
[*][MRC][MRC]	77	Memory is empty.
[1]	1	1 replaces 77 on display.
[=]	1994	

This is a trick, and it works this way: if you do not press 1, you will get 1994 × 77 = 153538 on the display. But because you are replacing 77 by 1, you get 1994 × 1 = 1994.

SQUARE ROOT KEY

Multiplying Numbers by Themselves

Keystrokes:	Display:
[2][*][2][=]	4
[3][*][3][=]	9
[4][*][4][=]	16
[10][*][10][=]	100
[138][*][138][=]	19044

There is a simpler way to multiply a number by itself. After entering a number, just press [*][=].

Keystrokes:	Display:
[2][*][=]	4
[3][*][=]	9
[4][*][=]	16
[10][*][=]	100
[138][*][=]	19044

A combination of keystrokes which does a task is called a program. So [*][=] is a program for multiplying a number by itself. [*][=] should always be used when you multiply a number by itself—not because it is simpler, but because entering a number a second time increases the chances of making an error.

Another way to say "multiply a number by itself" is "compute the square of a number." These two expressions mean the same thing. The first is better, but the second is used more often.

The keystroke sequence [*][=] should be read "times, equal." It should not be read "multiply by equal"! Nothing is being multiplied by equal; it makes no sense. [*][=] multiplies the number on the display by itself, and in order to execute the program, the times key and the equal key must be pressed.

Going Back

After you have multiplied a number by itself, pressing the [√] key restores the original number.

Restoring the original number is called computing the square root (the [√] key is called the square root key), because it reverses the operation of computing the square of a number.

Keystrokes:	Display:
[2][*][=][√]	2
[3][*][=][√]	3
[4][*][=][√]	4
[10][*][=][√]	10
[138][*][=][√]	138

Also notice that:

Keystrokes:	Display:
[4][√][*][=]	4
[9][√][*][=]	9

But this keystroke order does not always work.

(1) Problems with rounding errors:

Keystrokes:	Display:
[2][√][*][=]	1.9999998
[3][√][*][=]	2.9999999

The results are not exactly 2 and 3. These are errors made by the calculator. Because the square roots of 2 and 3 are represented only approximately as decimal fractions, a small error occurs.

(2) Problems with negative numbers:

Keystrokes:	Display:	
[4][+/−][*][=][√]	4	
[4][+/−][√][*][=]	E 2	(An error occurred.)

An error occurred in the second case, because the square root is not defined for negative numbers.

This is a difficult topic, because in addition to real numbers there is another system of numbers called complex numbers, in which you can compute square roots of negative numbers. We suggest that the topic of roots of negative numbers be avoided.

Applications

There are many applications of the square root in geometry. (See for example Chapter 33, titled "Computing Lengths of Diagonals with the Pythagorean Theorem.") They are mainly related to three topics: the Pythagorean theorem, the distance between points given by their coordinates, and constructing geometric figures having specified areas.

PLUS-MINUS KEY: COLD TEMPERATURES

See below for an activity involving temperature.
What does [+/−] (the plus-minus key) do?
It puts − in front of a number, or it removes −, if it was already there.
Examples:

Keystrokes:	Display:
[15]	15

[+/−]	−15
[+/−]	15
[+/−]	−15

(On the calculator that we are using, − appears in the left-most column of the display.)
Does it always work this way? Yes.

Try: [1][+/−][2][+/−][+/−][3][+/−],
[ON/C][1][+][=][=][=][+/−][=][=][+/−][=][=],
[2.5][*][2][=][+/−].

All numbers (except 0) that have a minus sign in front are called negative numbers. Numbers (except 0) without a minus sign are called positive numbers. Zero is neither positive nor negative, regardless of how it is written (0, or −0). You may say that 0 equals −0 (0 = −0).

You write:	You read:	Or you read:
−5	minus five	negative five
5	five	
−3.2	minus 3 point 2	negative 3 point 2

You saw that [5][+/−][+/−] gives you back 5. You may write it as − −5 = 5, and you may say the negative of negative five is five, or the negative of minus five is five.

How do you get negative numbers without using the [+/−] key? By subtracting a bigger number from a smaller number.

Try:	Keystrokes:	Display:
	[1][−][2][=]	−1
	[20][−][100][=]	−80
	[0][−][10][=]	−10

All negative numbers are smaller than 0.

Try: [1][+/−][−][0][=] (Display: −1)

When do you see negative numbers?
When it becomes very cold outside during winter.

Temperature Game

On the calculator we will keep the temperature. It is not the real temperature, but we will pretend that it is. Then we do one of two things:

(1) We say it has gotten cooler by some amount. For example, "The temperature has dropped five degrees." We see what has happened by pressing [−][5][=] (subtracting 5).
(2) Or we say it has gotten warmer by some amount. For example, "The temperature has risen eight degrees." We see what has happened by pressing [+][8][=].

When it becomes really cold we see a negative number on our calculator.

Example:

We say:	We press:	We see:	
	[50]	50	(Not so cold.)
The temperature dropped twenty degrees.	[−][20][=]	30	(It is freezing.)
The temperture dropped another twenty degrees.	[−][20][=]	10	(It is very cold.)
The temperature dropped fifteen degrees.	[−][15][=]	−5	(It is minus five degrees; better stay inside!)
The temperature rose twenty-three degrees.	[+][23][=]	18	(It is warmer.)

PERCENTAGE PROBLEMS

(1) Three people went on a raft trip. They decided to share the cost unequally. The first paid 50% of the total, the second paid 35%, and the third paid the remaining 15%. The total cost of the trip was $349. How much should each person pay?

Keystrokes:	Display:		
[349][*][50][%]	174.5	The first pays	$174.50
[35][%]	122.15	The second pays	$122.15
[15][%]	52.35	The third pays	$ 52.35

Notice that you do not have to reenter the amount when you are computing different percentages of the same total.
To check the answers, compute:

[174.5][+][122.15][+][52.35][=] Display: 349.

(2) Two people went to a restaurant. The check was $11.63. They wanted to leave a 14% tip and share the cost equally. How much should each of them pay?

Keystrokes:	Display:	
[11.63][+][14][%]	13.2582	Total plus tip.
[÷][2][=]	6.6291	Each person pays $6.63.

Note: when using [%], you do not press the [=] key.
It is very likely that the people would round $6.63, and they might do it differently: $6.50 or $6.75.

(3) A sweater was on sale. It was marked down 40% from its original price of $39.99. The sales tax is 6.5%. What is the total sale price?

Keystrokes:	Display:	
[39.99][−][40][%]	23.994	Sale price.
[+][6.5][%]	25.55361	The total price is $25.55.

Note: a store usually rounds the number after each computation. So the store would compute 6.5% tax on $23.99, and not on $23.994.

[23.99][+][6.5][%] gives 25.54935; the price is still $25.55.

(4) Jack and Jill were both earning $300 weekly. Jack's salary was cut 20%, but later he got a 20% raise. Jill got a 20% raise, but later she got a 20% salary cut. How much does each of them earn now? Are they earning the same amount?

Keystrokes:	Display:	
[300][−][20][%]	240	Jack's pay after cut.
[+][20][%]	288	Jack's pay after raise.
[300][+][20][%]	360	Jill's pay after raise.
[−][20][%]	288	Jill's pay after cut.

Their salaries are the same but lower than before. This is so because [−][20][%] is not really subtraction but multiplication, and it means [*][0.80][=]. (Here $0.80 = 1 - 0.20$.) And [+][20][%] means [*][1.20][=]. (Here $1.20 = 1 + 0.20$.)

Keystrokes:	Display:
[300][*][0.80][=]	240
[*][1.20][=]	288

So Jack's pay is $300 \times 0.80 \times 1.20 = 300 \times 0.96 = 288$ dollars, and Jill's pay is also $300 \times 1.20 \times 0.80 = 300 \times 0.96 = 288$ dollars.

(5) A company had to pay 37% tax on 60% of its capital gain. The capital gain was $365,847. How much was the tax?

Keystrokes:	Display:	
[365847][*][60][%]	219508.2	Taxable part of the profit.
[*][37][%]	81218.034	Amount of tax.

The company should pay $81,218. (Round the tax to the nearest dollar.)

Note: compare the program in Problem 5 to the program in Problem 1. In Problem 5 we compute 37% of the result of the previous computation, and not of the original amount, so the [*] key must be pressed again. (We press [*][37][%] and not just [37][%].)

APPENDIX